T0092955

Design of High Strength Steel Reinforced Concrete Columns

A Eurocode 4 Approach

Design of High Strength Steel Reinforced Concrete Columns

A Eurocode 4 Approach

Sing-Ping Chiew
Yan-Qing Cai

CRC Press
Taylor & Francis Group
Boca Raton London New York

CRC Press is an imprint of the
Taylor & Francis Group, an **informa** business

CRC Press
Taylor & Francis Group
6000 Broken Sound Parkway NW, Suite 300
Boca Raton, FL 33487-2742

© 2018 by Taylor & Francis Group, LLC
CRC Press is an imprint of Taylor & Francis Group, an Informa business

No claim to original U.S. Government works

Printed on acid-free paper

International Standard Book Number-13: 978-0-8153-8460-1 (Hardback)
978-1-138-60269-4 (Hardback)

Library of Congress Cataloging-in-Publication Data

Names: Chiew, Sing-Ping, author. | Cai, Y. Q. (Yan Qing), author.
Title: Design of high strength steel reinforced concrete columns : a Eurocode 4 approach / S.P. Chiew and Y.Q. Cai.
Description: Boca Raton : CRC Press, [2018] | Includes bibliographical references and index.
Identifiers: LCCN 2017057555 (print) | LCCN 2018000768 (ebook) | ISBN 9781351203944 (Adobe PDF) | ISBN 9781351203937 (ePub) | ISBN 9781351203920 (Mobipocket) | ISBN 9780815384601 (hardback : acid-free paper) | ISBN 9781351203951 (ebook)
Subjects: LCSH: Composite construction--Specifications--Europe. | Building, Iron and steel--Specifications--Europe. | Reinforced concrete construction--Specifications--Europe.
Classification: LCC TA664 (ebook) | LCC TA664 .C48 2018 (print) | DDC 624.1/83425--dc23
LC record available at https://lccn.loc.gov/2017057555

Visit the Taylor & Francis Web site at
http://www.taylorandfrancis.com

and the CRC Press Web site at
http://www.crcpress.com

Contents

List of symbols

A_a	Area of the structural steel
A_c	Area of concrete
A_{ch}	Area of highly confined concrete
A_{cp}	Area of partially confined concrete
A_{cu}	Area of unconfined concrete
A_s	Area of reinforcement
E_a	Modulus of elasticity of structural steel
$E_{c,eff}$	Effective modulus of elasticity of concrete
E_{cm}	Secant modulus of elasticity of concrete
$E_c(t)$	Tangent modulus of elasticity of concrete at time t
E_s	Modulus of elasticity of reinforcement
$(EI)_{eff}$	Effective flexural stiffness
G_a	Shear modulus of structural steel
I	Second moment of area of the composite section
I_a	Second moment of area of the structural steel
I_c	Second moment of area of the concrete
I_s	Second moment of area of the reinforcement
K_e	Correction factor
L	Length
M_{Ed}	Design bending moment
$M_{pl,a,Rd}$	The plastic resistance moment of the structural steel
$M_{pl,Rd}$	The plastic resistance moment of the composite section
N_{cr}	Elastic critical force in composite columns
N_{Ed}	The compressive normal force
$N_{pl,Rd}$	The plastic resistance of the composite section
$N_{pl,Rk}$	Characteristic value of the plastic resistance of the composite section
$N_{pm,Rd}$	The resistance of the concrete to compressive normal force
P_{Rd}	The resistance of per shear stud

V_{Ed}	The shear force
$V_{pl,a,Rd}$	The shear resistance of the steel section
W_{pa}	The plastic section modulus of the structural steel
W_{pc}	The plastic section modulus of the concrete
W_{ps}	The plastic section modulus of the reinforcing steel
b_c	Width of the composite section
b_f	Width of the steel flange
c_y, c_z	Thickness of concrete cover
d	Diameter of shank of the headed stud
e	Eccentricity of loading
f_{ck}	The cylinder compressive strength of concrete
f_{cd}	The design strength of concrete
$f_{c,p}$	The compressive strength of partially confined concrete
$f_{c,h}$	The compressive strength of highly confined concrete
f_s	The yield strength of reinforcement
f_u	Tensile strength
f_y	The yield strength of structural steel
f_{yd}	The design strength of structural steel
f_{yh}	The yield strength of transverse reinforcement
h_a	Depth of steel section
h_c	Depth of composite section
h_n	Distance from centroidal axis to neutral axis
h_{sc}	Overall nominal height of the headed stud
s	Spacing center-to-center of links
t_f	Thickness of steel flange
t_w	Thickness of the steel web
$\Delta\sigma$	Stress range
Ψ	Coefficient
α	Coefficient; factor
β	Factor; coefficient
γ	Partial factor
δ	Steel contribution ratio
η	Coefficient
$\varepsilon_{c,u}$	Strain of unconfined concrete
$\varepsilon_{c,p}$	Strain of partially confined concrete
$\varepsilon_{c,h}$	Strain of highly confined concrete
μ	Factors related to bending moments
λ	Relative slenderness
ρ_s	Reinforcement ratio
χ	Reduction factor of buckling
φ	Creep coefficient

Preface

This book is the companion volume to Design Examples for High-Strength Steel-Reinforced Concrete Columns—A Eurocode 4 Approach.

The current design codes for composite steel-reinforced concrete (SRC) columns are limited to normal strength steel and concrete materials. In Eurocode 4 (EC4) design, the range of concrete grades allowed is C20/25—C50/60 and steel grades allowed are S235—S460. However, in Eurocode 2 (EC2)—design of concrete structures, the grade of concrete allowed is up to C90/105 whereas in Eurocode 3 (EC3)—design of steel structures, the grade of structural steel allowed is up to S690.

SRC columns generally exhibit better ductility and higher buckling resistance when compared to either steel or reinforced concrete (RC) columns alone. However, EC4 gives a much narrower range of permitted concrete and steel material strengths in comparison to EC2 and EC3. This can be attributed to two main reasons: (1) lack of test data with high-strength materials, and (2) lack of strain compatibility between high-strength steel and concrete materials. In view of more recent research and understanding of SRC columns, the material strength limits in EC4 can be extended to match those in EC2 and EC3 to realize its full potential in composite steel concrete construction.

There is obviously a need for guidance on design of SRC columns using high-strength concrete, high-strength structural steel and high-strength-reinforcing steel materials. This book fills this gap by allowing the design of such columns with concrete cylinder strength up to 90 N/mm^2, yield strength of structural steel up to 690 N/mm^2 and yield strength of reinforcing steel up to 600 N/mm^2, respectively. The design is essentially based on EC4 approach with special considerations given to resistance calculations which maximize the full strength of the concrete, structural steel and reinforcing steel materials. The design methods are based on taking advantage of either appropriate concrete confinement model or creep and shrinkage model permitted by the design codes. Fire protection and fire-resistance design are also included for completeness.

This book is written primarily for structural engineers and designers who are familiar with basic EC4 design and wish to take advantage of higher-strength materials to design more competitive SRC columns safely and efficiently. Equations for design resistances are presented clearly so that they can be easily programmed into design spreadsheets for ease of use. Design procedures are demonstrated through examples worked out in the companion volume to this book, Design Examples for High Strength Steel Reinforced Concrete Columns—A Eurocode 4 Approach. This book should also be useful to civil engineering students who are studying composite steel concrete design and construction.

Authors

Sing-Ping Chiew is a professor and the Civil Engineering Programme director at the Singapore Institute of Technology, Singapore, and coauthor of *Structural Steelwork Design to Limit State Theory, 4th Edition.*

Yan-Qing Cai is a project officer in the School of Civil and Environmental Engineering at Nanyang Technological University, Singapore.

Chapter 1

Introduction

1.1 STEEL-REINFORCED CONCRETE COLUMNS

A steel-concrete composite column is a composite member subjected primarily to compression and bending. For a composite column, the steel and concrete act simultaneously in resisting the imposed loads through the bonding and friction between their interfaces. In the past few decades, composite columns have been widely used, as they combine the advantages of steel and concrete materials to achieve an overall enhancement in strength and stiffness.

Composite columns are classified into two principal categories: (1) steel-reinforced concrete sections and (2) concrete-filled hollow steel sections. Typical composite cross-sections of SRC columns are shown in Figure 1.1. In fully encased sections, the steel section is embedded within concrete, as shown in Figure 1.1a. Partially encased sections consist of steel I- or H-sections and the concrete filled in the void between the steel flanges, as shown in Figure 1.1b and c.

The typical construction process of SRC columns is shown in Figure 1.2. The internal steel section is erected first, followed by the placement of reinforcements and formwork, and thereafter concrete is poured into the formwork to form the SRC column.

SRC columns have been widely used in many countries due to various advantages, that is, high strength and stiffness and convenient construction. Generally, SRC columns provide better fire performance in comparison with pure steel columns and concrete filled steel tubes (CFT) columns because the external concrete cover provides protection for the steel section. Investigations on the fire performance of SRC columns have been conducted by many researchers. It was reported that the fire resistance increased as the concrete cover thickness increased.

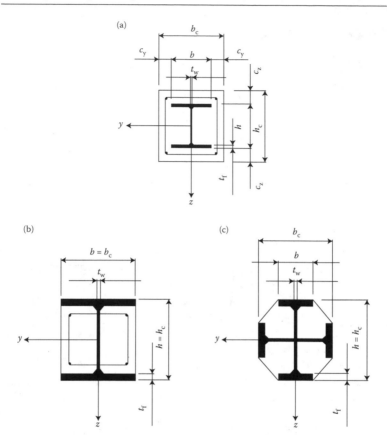

Figure 1.1 Typical cross-sections of SRC columns.

An additional advantage of SRC columns is that the concrete prevents the occurrence of local buckling (Weng and Yen, 2002).

1.2 APPLICATION OF HIGH-STRENGTH MATERIALS

Column construction consumes a considerable amount of resources, particularly in tall buildings and bridges. This is when steel-concrete composite columns are often employed to reduce the column size. Currently, composite columns are constructed using normal-strength steel (NSS) and normal-strength concrete (NSC). Incrementing the column capacity by using higher-strength materials can lead to many economic

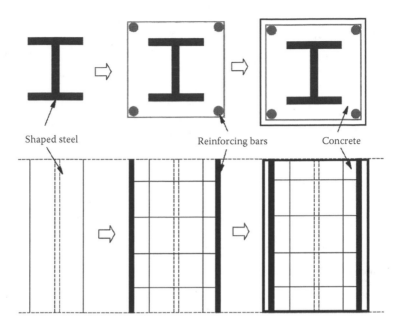

Figure 1.2 Construction process of SRC columns.

benefits for both the builders (reduction in materials and construction cost) and the users of the building (more useable area). Therefore, high-strength construction materials provide a cost-effective alternative to traditional materials. Their various advantages are listed as follows:

1. Provide higher strength and lower weight
2. Reduce member size
3. Free up more useable area
4. Require fewer construction materials and less construction work
5. Reduce the use of space and labor in handling materials

However, the ductility of material decreases with an increase in strength. Material brittleness is a major problem for high-strength concrete (HSC), as shown in Figure 1.3. Local buckling is a major problem for high-strength steel (HSS). To overcome these problems, SRC columns came into existence, where the ductility of concrete can be improved and the local buckling of steel sections can be prevented or minimized.

Today, steel grades up to S460 and concrete classes up to C50/60 have been widely used in construction all over the world. Higher steel grades

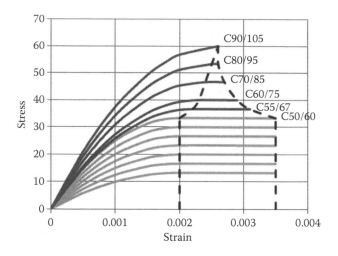

Figure 1.3 Stress–strain curves for different concrete classes.

(e.g., S690) and high-strength concretes (e.g., C90/105) have also been available for structural application for more than three decades. However, the use of high-strength materials is still limited in SRC columns due to the lack of design rules and inadequate field experience.

Today, high-strength materials have been used in seismically active countries, for example, the United States, Japan, Korea, and China. As material research and manufacturing technology improve, the use of higher-strength concrete and steel materials will continue to increase as wider applications are being sought in the construction of modern cities.

1.3 LIMITATION OF CURRENT DESIGN CODES

Current design codes for composite columns are limited to the application of normal-strength materials, as shown in Table 1.1. AISC 360-10 allows the usage of cylinder strength of normal-weight concrete in a range of 21–70 N/mm² and yield strength of steel up to 525 N/mm² in composite columns. Eurocode 4 (EC4) allows the usage of concrete classes C20/25–C50/60 and steel grades S235–S460 in composite columns. However, EC2 for concrete structure design allows the use of concrete classes up to C90/105, and EC3 for steel structure design allows the use of steel grades up to S690.

SRC columns generally exhibit better ductility or higher buckling resistance as compared to steel columns and reinforced concrete (RC)

Table 1.1 Limitations of material strength in current design codes

Codes	Yield strength of steel (N/mm²)	Cylinder strength of concrete (N/mm²)
AISC 360-10	≤525	21~70
EC4	235~460	20~50
EC2	–	≤90
EC3	≤700	–

columns. However, EC4 gives a narrower range of material strength of steel and concrete in comparison to EC2 and EC3. It could be mostly due to the following two factors:

1. *Lack of test data on SRC columns with high-strength materials*: In view of previous tests on SRC columns, many of the tests focus on SRC columns with normal-strength concrete and steel (with only a few on high-strength material). Therefore, there is insufficient test data to strongly establish the validity of using high-strength materials in SRC columns in EC4.
2. *Strain compatibility issues between high-strength steel and concrete*: Figure 1.4 shows why the strengths of NSS and NSC are utilized effectively: the yield strain of normal-strength steel

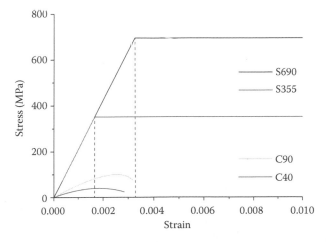

Figure 1.4 Stress–strain curves of steel and concrete.

is close to the peak strain of concrete. These two materials are therefore strain compatible and reach their full strengths when the composite section reaches the ultimate strength. The same plot shows that HSS and HSC are not strain compatible in the sense that HSC reaches its peak strain much earlier than the yield strain of HSS. This implies that concrete will fail earlier than steel and results in a partial utilization of the capacities of both the materials. For CFT, such a strain incompatibility is prevented by the confinement of the exterior steel section. The more ductile exterior steel section provides confinement that increases both the peak strain and strength of enclosed concrete. For SRC columns, there are four ways to increase the compressive strain: (1) use a new type of concrete Engineered Cementitious Composite (ECC) as an exterior skin of the concrete core, (2) take advantage of confinement from lateral hoops and steel section, (3) take advantage of creep and shrinkage effects, and (4) use external fiber-reinforced polymer. The most common approach is to use extra lateral reinforcement to provide adequate confinement for concrete. To ensure that the strength of high-strength steel can be fully developed, the strain compatibility problem should be considered in the column design.

1.4 ABOUT THIS BOOK

A series of design guides on SRC members have been produced by American, European, and Japanese authors; however, they do not provide guidance on the design of SRC members using high-strength concrete, high-strength steel, and high-strength reinforcing steel. This book fills the gap by examining the design of SRC members with concrete cylinder strength up to 90 N/mm^2, yield strength of structural steel up to 690 N/mm^2, and yield strength of reinforcing steel up to 600 N/mm^2. The design is based on EC4 with special consideration for high-strength concrete, high-strength reinforcing steel, and structural steel. The design method has been proposed based on the latest research data collected from the literature.

The design guide also gives details of confinement concrete models and the creep and shrinkage method for SRC member design. Fire protection and fire-resistant designs are introduced for SRC columns. This design guide will endow structural engineers with the confidence to use high-strength materials in a safe and efficient manner in design and construction.

Chapter 2

Materials

2.1 CONCRETE

British Standard European Norm (BS EN) 206-1 defines HSC as concrete with a compressive strength class higher than C50/60 in the case of normal-weight concrete and LC50/55 in the case of lightweight concrete. Strength classes for normal-weight concrete are defined as Cx/y, where x and y are the cylinder and cube compressive strength, respectively. For high-strength concrete, special care is required for production and testing, and special structural design requirements may be needed.

The strength classes for NSC and HSC used for the design of SRC columns are shown in Tables 2.1 and 2.2, respectively.

2.2 STRUCTURAL STEEL

Classification of steel materials is necessary to determine whether these materials should be allowed for structural use in the construction industry with or without any restrictions. The adequacy and reliability of steel materials should be verified against the material performance requirements as well as the quality assurance requirements, respectively, in the entire process of classification.

Certification is the process of rigorous evaluation of the specifications given in the British/European, American, Japanese, Australian/New Zealand, and Chinese material standards against the essential material performance requirements. The purpose of certification is to derive lists of certified steel materials. Only those materials with specifications that comply with the relevant material performance requirements are included in the list.

Table 2.1 Material properties of NSC

Properties of concrete	Concrete class						
	C20/25	C25/30	C30/37	C35/45	C40/50	C45/55	C50/60
f_{ck} (MPa)	20	25	30	35	40	45	50
$f_{ck,cube}$ (MPa)	25	30	37	45	50	55	60
f_{ctm} (MPa)	2.2	2.6	2.9	3.2	3.5	3.8	4.1
E_{cm} (GPa)	29	31	33	34	35	36	37

Table 2.2 Material properties of HSC

Properties of concrete	Concrete class				
	C55/67	C60/75	C70/85	C80/95	C90/105
f_{ck} (MPa)	55	60	70	80	90
$f_{ck,cube}$ (MPa)	67	75	85	95	105
f_{ctm} (MPa)	4.2	4.4	4.6	4.8	5.0
E_{cm} (GPa)	38	39	41	42	44

Based on verification against material performance and quality assurance requirements, steel materials are classified into three classes: Class 1, Class 2, and Class 3. Class 1 steel materials are certified steel materials manufactured with approved quality assurance. Two types of steel material are defined as Class 2. First, Class 2 steel materials are noncertified steel materials that meet the material performance requirements through material testing and are manufactured with approved quality assurance. Additionally, certified steel materials that fail in reliability assessment may be also treated as Class 2 steel materials on a case-by-case basis subject to approval by Building and Construction Authority (BCA) if their reliability can alternatively be guaranteed through rigorous material control and testing plans on site. Class 3 steel materials are steel materials that do not meet at least one of the two requirements: material performance and quality assurance requirements.

EC3 defines high-strength steel as structural steel with yield strength greater than 460 N/mm². Structural steel used in Singapore should be in compliance with BC1:2012: The Design Guide on the Use of Alternative Structural Steel to BS 5950 and Eurocode 3. The yield strength for normal-strength steel and HSS used for SRC columns is shown in Tables 2.3 through 2.7. The density of structural steel is assumed to be 7850 kg/m³. The modulus of elasticity of steel is taken as 210 GPa.

Table 2.3 Nominal values of yield strength of structural steel

	Yield strength f_y (MPa) for nominal thickness (mm) less than or equal to					
Grade	16	40	63	80	100	150
S235	235	225	215	215	215	195
S275	275	265	255	245	235	225
S355	355	345	335	325	315	295
S420	420	400	390	370	360	340
S460	460	440	430	410	400	380
S500	500	500	480	480	480	440
S550	550	550	530	530	530	490
S620	620	620	580	580	580	560
S690	690	690	650	650	650	630

Table 2.4 Design parameters of American (ASTM and API) structural steels

	Yield strength f_y (MPa) for thickness (mm) less than or equal to				
Grade	32	50	65	80	100
36 [250]	250	240	230	220	210
42 [290]	290	280	270	260	250
50 [345]	345	335	325	315	305
55 [380]	380	370	360	350	340
60 [415]	415	405	395	385	375
65 [450]	450	440	430	420	410

Table 2.5 Design parameters of Japanese (JIS) structural steels

	Yield strength f_y (MPa) for thickness (mm) less than or equal to					
Grade	16	40	75	100	160	200
400	245	235	215	215	205	195
490	325	315	295	295	285	275
490Y	365	355	335	325	–	–
520	365	355	335	325	–	–
570	460	450	430	420	–	–

Table 2.6 Design parameters of Australian/New Zealand (AS/NZS) structural steels

Grade	Yield strength f_y (MPa) for thickness (mm) less than or equal to					
	12	20	32	50	80	150
250	250	250	250	250	240	230
300	300	300	280	280	270	260
350	350	350	340	340	340	330
400	400	380	360	360	360	–
450	450	450	420	400	–	–
CA220	210	–	–	–	–	–
CA260	250	–	–	–	–	–
CA350	350	–	–	–	–	–
PT430	300	280	280	270	270	250
PT460	305	295	295	275	275	265
PT490	360	340	340	330	330	320
PT540	450	450	420	400	–	–

Table 2.7 Design parameters of Chinese (GB) structural steels

Grade	Yield strength f_y (MPa) for thickness (mm) less than or equal to				
	16	35	50	100	150
Q235	235	225	215	215	195
Q275	275	265	255	245	225
Q295	295	275	255	235	–
Q345	345	325	295	275	–
Q355	355	345	335	325	–
Q390	390	370	350	330	–
Q420	420	400	380	360	–
Q460	460	440	420	400	–

The limiting values of the ratio f_u/f_y, the elongation at failure, and the ultimate strain ε_{cu} for normal-strength and high-strength steel are recommended in Table 2.8.

2.3 REINFORCING STEEL

The yield strength of reinforcing steel is limited to the range of 400–600 MPa in accordance with EN1992-1-1. Reinforcing steel used in

Table 2.8 Limitations on tensile ratio, elongation, and ultimate
strain for structural steel

	Normal-strength steel	High-strength steel
Ratio f_u/f_y	≥ 1.10	≥ 1.05
Elongation at failure	$\geq 15\%$	$\geq 10\%$
Ultimate strain	$\varepsilon_{cu} \geq 15\, f_y/E$	$\varepsilon_{cu} \geq 15\, f_y/E$

Table 2.9 Mechanical properties of reinforcing steel

	Yield strength (MPa)	Tensile/yield strength ratio	Elongation at maximum force %
B500A	500	1.05[a]	2.5[b]
B500B	500	1.08	5.0
B500C	500	$\geq 1.15, <1.35$	7.5
B600A	600	1.05[a]	2.5[b]
B600B	600	1.08	5.0
B600C	600	$\geq 1.15, <1.35$	7.5

[a] Tensile/yield strength ratio is 1.02 for sizes below 8 mm.
[b] Elongation is 1.0% for sizes below 8 mm.

Singapore should be in compliance with SS560:2016: Specification for Steel for the Reinforcement of Concrete—Weldable Reinforcement Steel—Bar, Coil and Decoiled Product. The mechanical properties of reinforcing steel are shown in Table 2.9. For composite members, the modulus of elasticity of reinforcement is taken as 210 GPa, which is equal to the value of structural steel, rather than 200 GPa in reinforced concrete structures.

2.4 SHEAR CONNECTORS

Shear connectors are used for providing composite action between steel and concrete. This connection (referred to as shear connection) is provided mainly to resist longitudinal shear action. The most common connector used in construction is the headed shear connector. It can be welded to the upper flange of steel beams either directly or through profiled steel sheeting. Figure 2.1 shows a typical headed shear connector.

The shank diameter should be in the range of 10–25 mm. The overall height of a stud should be at least three times the shank diameter. The head diameter should be at least 1.5 times the shank diameter, whereas the head depth should be at least 0.4 times the shank diameter. The nominal tensile strength should be at least 400 MPa.

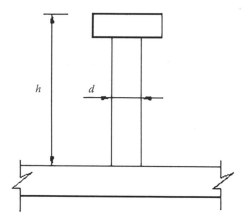

Figure 2.1 Headed shear connector.

Stud shear connectors used in Singapore should be in compliance with BC1:2012: The Design Guide on the Use of Alternative Structural Steel to BS 5950 and Eurocode 3. The tensile strengths of stud shear connectors manufactured for EN, ASTM, JIS, AS/NZS, and GB are given in Table 2.10.

Shear connectors other than the headed stud connectors are allowed, provided they can perform in accordance with the product manufacturer's recommendations.

The design equations given in EN 1994-1-1 are based on some factors, such as shank diameter d and ultimate strength f_u of headed studs, characteristic strength f_{ck} and modulus E_{cm} of concrete, and failure mode either in the stud shear connector or the concrete. The design shear resistance of a headed stud automatically welded in accordance with EN14555 should be determined from:

Table 2.10 Tensile strengths of British/European
(BS EN), American (AWS), Japanese
(JIS), Australian/New Zealand (AS/NZS),
and Chinese (GB) stud shear connectors

Material standards	Tensile strength f_u (MPa)
BS EN ISO 13918	500
AWS D1.1 (Type B)	450
JIS B 1198	400
AS/NZS 1554.2	410
GB/T 10433	400

$$P_{Rd} = \frac{0.8 f_u \pi d^2/4}{\gamma_V} \tag{2.1}$$

or

$$P_{Rd} = \frac{0.29\alpha d^2 \sqrt{f_{ck} E_{cm}}}{\gamma_V} \tag{2.2}$$

whichever is smaller, with:

$$\alpha = 0.2(h_{sc}/d + 1) \quad \text{for } 3 \leq h_{sc}/d \leq 4 \tag{2.3}$$

$$\alpha = 1.0 \quad \text{for } h_{sc}/d > 4 \tag{2.4}$$

where
   γ_V: is the partial factor; take as 1.25;
   d: is the diameter of the shank of the headed stud, 16 mm $\leq d \leq$ 25 mm;
   f_u: is the ultimate tensile strength of the material of the headed stud
      ($f_u \leq$ 500 N/mm^2);
   f_{ck}: is the characteristic cylinder compressive strength of concrete;
   h_{sc}: is the overall nominal height of the headed stud; and
   E_{cm}: is the modulus of concrete.

Equation 2.1 is based on failure of the shank of the headed stud and Equation 2.2 is based on failure in concrete. The lower of the above two values governs the design of shear connectors. Accordingly, when high-strength concrete is used, the design resistance will be governed by the failure of the shear connectors. The design resistance values of some commonly used shear connectors are given in Appendix A.

2.5 TEST DATABASE ON STEEL-REINFORCED CONCRETE COLUMNS

Experimental investigation on SRC columns has been conducted by Anslijn and Janss (1974), Stevens (1965), Matsui (1979), SSRC Task Group 20 (1979), Roik and Diekmann (1989), Han et al. (1992), Mirza et al. (1996), Chen and Yeh (1996), Chen and Lin (2006), Hoang (2009), Kim et al. (2012), and so on. These tests were conducted on SRC columns with various concrete and steel strengths, steel sections, reinforcement ratios, slenderness ratios, and eccentricity ratios. The dimension and material properties of test specimens are summarized in Table 2.11, where $b \times h$ is the cross-section of SRC columns; f_c is the compressive cylinder

Table 2.11 Summary of dimensions and material properties of SRC test

Properties	$b \times h$ (mm × mm)	f_c (MPa)	f_y (MPa)	f_s (MPa)	ρ_{ss} (%)	ρ_s (%)	e/h
Minimum value	160 × 160	13	223	276	2.7	0	0
Maximum value	406 × 305	104	913	642	15	3.4	1.5

strength of concrete; f_y and f_s are the yield strength of structural steel and reinforcement, respectively; ρ_{ss} is the steel ratio; ρ_s is the reinforcement ratio; and *e/h* is the eccentricity ratio.

Figures 2.2 and 2.3 illustrate the distribution of steel and concrete strength in earlier tests, respectively. They indicate relatively fewer SRC column tests with high-strength steel and concrete. Therefore, test data are insufficient to establish the validity of using high-strength materials in SRC columns in EC4.

In this database, test specimens involve short and long SRC members subjected to compression and bending. The test results are used for comparison with the resistance predicted by EC4 to establish the design guide for using high-strength materials for SRC columns.

The test/EC4 ratios against concrete compressive cylinder strength are shown in Figure 2.4. The ratios are categorized into two groups based on concrete compressive cylinder strength. For normal-strength concrete ($f_{ck} < 50$ MPa), the test/EC4 value for SRC specimens is greater than unity, generally indicating a conservative average prediction by EC4. For high-strength concrete ($f_{ck} > 50$ MPa), the test/EC4 value for SRC

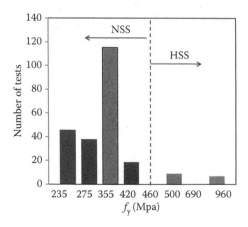

Figure 2.2 Distribution of f_y.

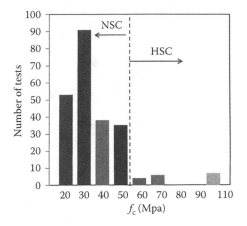

Figure 2.3 Distribution of f_c.

members is less than unity, generally indicating an unconservative prediction by EC4.

The test/EC4 ratios against steel yield strength are shown in Figure 2.5. The ratios are categorized into two groups based on steel strength. For NSS ($f_y \leq 460\text{ N/mm}^2$), the test/EC4 value for SRC specimens is greater than unity, generally indicating a conservative average prediction by EC4. However, this is not true for SRC columns with HSS ($f_y > 460\text{ N/mm}^2$). The test/EC4 value for SRC members with HSS is less than unity, generally indicating an unconservative prediction

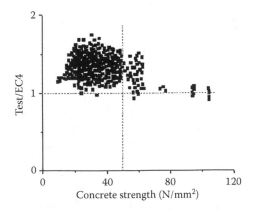

Figure 2.4 Comparison of test/EC4 ratio against concrete cylinder strength.

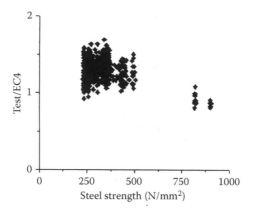

Figure 2.5 Comparison of test/EC4 ratio against steel strength.

by EC4. This might be due to the strain-compatibility issue between concrete and steel. Thus, for the use of HSS in SRC, the liability needs to be further investigated.

As the test data are insufficient to establish the validity of using high-strength materials in accordance with EC4, the following section provides additional guidelines for the use of HSS in SRC columns by considering the strain-compatibility and confinement effects from lateral hoops for SRC columns.

2.6 STRAIN COMPATIBILITY

For NSS and NSC, the yield strain of steel is very close to the peak strain of concrete. The two materials are therefore strain-compatible and able to reach their full strengths when the composite concrete section reaches its ultimate strength. High-strength steel and concrete are not strain-compatible in the sense that HSC reaches its peak strain much earlier than the yield strain of high strength steel. This implies that concrete will fail earlier than steel, resulting in a partial utilization of the capacities of the materials. The structural steel cannot develop its full strength, so the plastic method provided by EC4 is expected to overestimate the resistance of SRC columns with high-strength steel. In this sense, the strain-compatibility method seems to be the more appropriate method for predicting the resistance of SRC columns. Using the strain-compatibility

Table 2.12 Maximum strength of steel in SRC columns

	S235	S275	S355	S420	S460	S500	S550	S620	S690
C20/25	235	275	355	413	413	413	413	413	413
C25/30	235	275	355	420	435	435	435	435	435
C30/37	235	275	355	420	454	454	454	454	454
C35/45	235	275	355	420	460	472	472	472	472
C40/50	235	275	355	420	460	488	488	488	488
C45/55	235	275	355	420	460	500	503	503	503
C50/60	235	275	355	420	460	500	518	518	518
C55/67	235	275	355	420	460	500	531	531	531
C60/75	235	275	355	420	460	500	544	544	544
C70/85	235	275	355	420	460	500	550	567	567
C80/95	235	275	355	420	460	500	550	589	589
C90/105	235	275	355	420	460	500	550	589	589

method, the maximum strength of structural steel that can be utilized in SRC can be derived.

The yield strain of steel ε_y and the peak strain of concrete ε_c can be calculated in accordance with EC2 and EC3. To ensure that the yield strain of steel is less than the compressive strain of concrete, that is, $\varepsilon_y \leq \varepsilon_c$, the maximum utilizable strength of steel can be determined accordingly. Table 2.12 gives the maximum steel strength in SRC columns such that $\varepsilon_y < \varepsilon_c$. Based on the recommendation for CFT from

Table 2.13 Compatibility of steel and concrete materials for SRC columns

	S235	S275	S355	S420	S460	S500	S550	S620	S690
C20/25	Yes	Yes	Yes	No	No	No	No	No	No
C25/30	Yes	Yes	Yes	Yes	No	No	No	No	No
C30/37	Yes	Yes	Yes	Yes	No	No	No	No	No
C35/45	Yes	Yes	Yes	Yes	Yes	No	No	No	No
C40/50	Yes	Yes	Yes	Yes	Yes	No	No	No	No
C45/55	Yes	Yes	Yes	Yes	Yes	Yes	No	No	No
C50/60	Yes	Yes	Yes	Yes	Yes	Yes	No	No	No
C55/67	Yes	Yes	Yes	Yes	Yes	Yes	No	No	No
C60/75	Yes	Yes	Yes	Yes	Yes	Yes	No	No	No
C70/85	Yes	Yes	Yes	Yes	Yes	Yes	Yes	No	No
C80/95	Yes	Yes	Yes	Yes	Yes	Yes	Yes	No	No
C90/105	Yes	Yes	Yes	Yes	Yes	Yes	Yes	No	No

Note: "Yes" indicates compatible materials and "No" is not recommended.

Liew and Xiong (2015), a recommendation on the matching grades of steel and concrete suitable for use in SRC columns was proposed as shown in Table 2.13. It is recommended that concrete classes up to C90/105 be used with steel grades up to S550 in SRC columns. The above calculation ignores the confinement effect from the links and steel sections. If the increase of strain by confinement is taken into account, higher steel grades could be used in SRC columns.

For SRC columns, the lateral hoops and steel sections can provide lateral confining pressure for the concrete core, which results in an enhancement in the strength and ductility of the concrete, depending on the degree of the confining pressure. In addition, confinement is also affected by other factors, such as the layout of longitudinal reinforcement, cross-section configuration, and loading type. With the increased maximum strain of concrete, the full strength of the HSS section may be developed.

Chapter 3

Concrete confinement model

3.1 GENERAL

The strength and ductility of concrete were enhanced by the confining pressure from lateral hoops and steel sections based on earlier research works. It has been shown that the behavior of the confinement is dependent on various factors such as the shape of the structural steel, the diameter, layout, spacing and number of longitudinal reinforcements and transverse rebar, and the yield strength of the rebar, as well as the concrete strength. According to Mirza and Skrabek (1992), concrete in SRC columns can be divided into three regions: the unconfined concrete (UCC) region, partially confined concrete (PCC) region within the hoops, and highly confined concrete (HCC) region between the flanges of the steel section (Figure 3.1).

For PCC, the confinement effect can be determined from the expressions for reinforced concrete columns, which were extensively studied by Mander et al. (1988), Cusson and Paultre (1995), and Legeron and Paultre (2003). Additionally, CEB Model Code 1990, EC2, and FIB Model Code 2010 also provide expressions for calculating the stress–strain of confined concrete in RC columns. Thus, an accurate confinement model can be identified from the models given in researcher papers or design codes.

For highly confined concrete, an analytical model was proposed by Chen and Lin (2006) to predict the behavior of SRC columns. The ranges of two factors (partial confinement factor and high confinement factor) are given in accordance with several experimental tests. The range of the high confinement factor is 1.1–1.97. The range of the partial confinement factor is 1.09–1.5. However, the results are case dependent, and no interaction mechanism between concrete and steel sections or formula has been developed.

The modified confinement model for SRC columns can be developed by combining the verified PCC model and the HCC model.

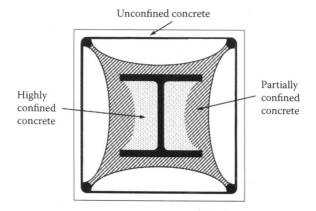

Figure 3.1 Confinement zones in SRC columns.

3.2 CONFINEMENT MODEL FOR PARTIALLY CONFINED CONCRETE (REINFORCED CONCRETE MEMBERS)

3.2.1 Mander model

A stress–strain model for concrete confined by different shapes of transverse reinforcement was proposed by Mander et al. (1988). The stress–strain relationship is shown in Figure 3.2. According to the Mander model, the compressive strength of confined concrete f'_{cc} is determined from the cylinder strength of unconfined concrete f'_{co} and the lateral confining pressure f'_l:

$$f'_{cc} = f'_{co}\left(-1.254 + 2.254\sqrt{1 + \frac{7.94 f'_l}{f'_{co}}} - 2\frac{f'_l}{f'_{co}}\right) \quad (3.1)$$

The peak strain at peak stress is given by

$$\varepsilon_{cc} = \varepsilon_{co}\left[1 + 5\left(\frac{f'_{cc}}{f'_{co}} - 1\right)\right] \quad (3.2)$$

where f'_{cc} and f'_{co} are the compressive strength of confined and unconfined concrete, respectively; f'_l is the effective lateral confining stress; and ε_{cc} and ε_{co} are the strain of confined and unconfined concrete, respectively.

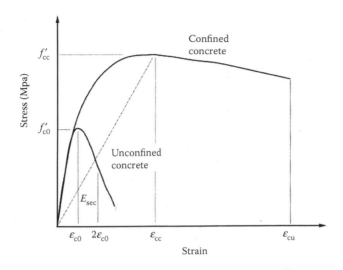

Figure 3.2 Stress–strain relationship from the Mander model.

The lateral confining pressure depends on the strength and volume ratio of the transverse reinforcements, layout of reinforcing steel, and confined concrete area.

As shown in Figure 3.3, arching action occurs horizontally between longitudinal reinforcements and vertically between layers of transverse

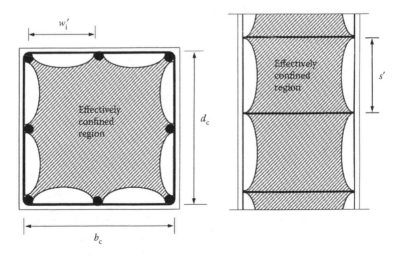

Figure 3.3 Effectively confined region.

reinforcement. The confinement effective coefficient is calculated by the following:

$$k_e = \frac{\left[1 - \sum_{i=1}^{n} \frac{(w_i')^2}{6b_c d_c}\right]\left(1 - \frac{s'}{2b_c}\right)\left(1 - \frac{s'}{2d_c}\right)}{1 - \rho_{cc}} \qquad (3.3)$$

where b_c and d_c are the width and depth of confined concrete, respectively; s' is the spacing between lateral hoops; ρ_{cc} is the volumetric ratio of longitudinal reinforcement; and w_i' is the distance between adjacent longitudinal reinforcements.

The ratio of lateral hoops to confined concrete in the y and z directions is

$$\rho_y = \frac{A_{sy}}{s'd_c} \qquad (3.4)$$

$$\rho_z = \frac{A_{sz}}{s'b_c} \qquad (3.5)$$

where A_{sy} and A_{sz} are the lateral reinforcement area in corresponding directions.

The lateral confining pressure in the y and z directions is

$$f_{ly} = \rho_y f_{yh} \qquad (3.6)$$

$$f_{lz} = \rho_z f_{yh} \qquad (3.7)$$

The effective lateral confining is

$$f_l' = k_e f_l \qquad (3.8)$$

where f_l is the lateral pressure from the transverse reinforcement; and f_{yh} is the yield strength of the transverse reinforcement.

The Mander model assumes the yielding of hoops at the peak strain of confined concrete. However, experimental studies conducted by Sheikh

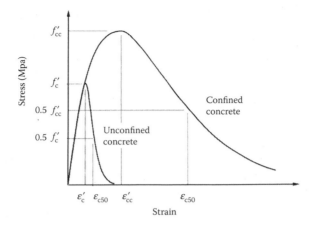

Figure 3.4 Stress–strain curve from the Legeron model.

and Uzumeri (1980) revealed that high-strength hoops may not yield when the confinement effect is small. Thus, modification is needed to calculate the real stress for hoops. A procedure proposed by Legeron and Paultre (2003) was adopted herein.

3.2.2 Legeron and Paultre model

A confinement model for RC members was defined by Cusson and Paultre (1995). Legeron and Paultre (2003) modified this model based on strain compatibility, as shown in Figure 3.4. The curve represents two critical points: (1) the maximum compressive strength f'_{cc} and strain ε'_{cc}; and (2) compressive strain ε_{cc50} corresponding to 0.5 f'_{cc}.

The maximum compressive stress and strain of confined concrete are given by:

$$\frac{f'_{cc}}{f'_c} = 1 + 2.4\left(I'_e\right)^{0.7} \tag{3.9}$$

$$\frac{\varepsilon'_{cc}}{\varepsilon'_c} = 1 + 35\left(I'_e\right)^{1.2} \tag{3.10}$$

With ε_{cc50}

$$I'_e = \frac{f'_{le}}{f'_c} \tag{3.11}$$

$$f'_{le} = \rho_{se} f'_h \tag{3.12}$$

$$\rho_{se} = k_e \rho_s \tag{3.13}$$

where f'_c and ε'_c are the compressive strength and strain of unconfined concrete; k_e is the confinement coefficient introduced by Mander et al. (1988); I_{e50} is the effective confinement index; and ρ_s is the volumetric ratio of the hoops.

The stress of confining steel f'_h is given as:

$$f'_h = \begin{cases} f_{yh} & \kappa \leq 10 \\ \dfrac{0.25 f'_c}{\rho_{se}(\kappa - 10)} \geq 0.43 \varepsilon'_c E_s & \kappa \geq 10 \end{cases} \tag{3.14}$$

With

$$\kappa = \frac{f'_c}{\rho_{se} E_s \varepsilon'_c} \tag{3.15}$$

where E_s is the elastic modulus of reinforcement.

The postpeak strain ε_{cc50} corresponding to 50% f'_{cc} is defined as

$$\frac{\varepsilon_{cc50}}{\varepsilon_{c50}} = 1 + 60 I_{e50} \tag{3.16}$$

With

$$I_{e50} = \rho_{se} \frac{f_h}{f'_c} \tag{3.17}$$

where ε_{c50} is the postpeak strain of unconfined concrete measured at 0.5 f'_c.

3.2.3 Eurocode2 model (European Committee for Concrete model code 90)

EC2 provides the stress–strain relationship of confined concrete in RC members, as shown in Figure 3.5. The characteristic strength and strain of confined concrete are determined from:

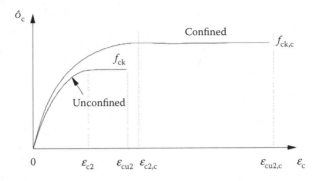

Figure 3.5 Stress–strain curve of confined concrete (EC2).

$$f_{ck,c} = f_{ck}(1.0 + 5.0\sigma_2/f_{ck}) \quad \text{for } \sigma_2 \leq 0.05 f_{ck}$$
$$f_{ck,c} = f_{ck}(1.125 + 2.5\sigma_2/f_{ck}) \quad \text{for } \sigma_2 > 0.05 f_{ck} \tag{3.18}$$

$$\varepsilon_{c2,c} = \varepsilon_{c2}(f_{ck,c}/f_{ck})^2 \tag{3.19}$$

$$\varepsilon_{cu2,c} = \varepsilon_{cu2} + 0.2\sigma_2/f_{ck} \tag{3.20}$$

where σ_2 is the lateral pressure due to confinement; ε_{c2} and $\varepsilon_{c2,c}$ are the strain of unconfined and confined concrete at peak stress, respectively; and ε_{cu2} and $\varepsilon_{cu2,c}$ are the ultimate strain of unconfined and confined concrete, respectively.

The effective lateral compressive stress due to confinement can be determined according to EC8:

$$\sigma_2/f_{ck} = 0.5\alpha\omega_{wd} \tag{3.21}$$

where
ω_{wd} is the volumetric ratio of the lateral hoops:

$$\omega_{wd} = \frac{\text{Volume of hoops}}{\text{Volume of confined concrete}} \frac{f_{yhd}}{f_{cd}}$$

α is the confinement factor, $\alpha = \alpha_n\alpha_s$, with:

$$\alpha_n = 1 - \frac{\sum b_i^2/6}{b_0 h_0} \tag{3.22}$$

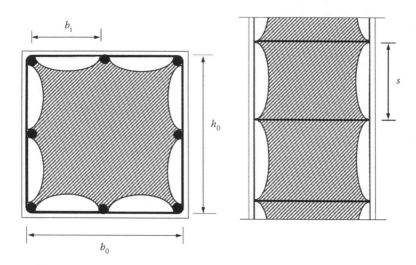

Figure 3.6 Dimensions used in determining the confining effect.

$$\alpha_s = \left(1 - \frac{s}{2b_0}\right)\left(1 - \frac{s}{2h_0}\right) \tag{3.23}$$

where f_{yhd} is the design strength of the transverse reinforcement; f_{cd} is the design strength of concrete; n is the total number of longitudinal reinforcements laterally engaged by hoops; b_i is the distance between consecutive engaged bars; h_0 is the depth of the confined core; b_0 is the width of the confined core; and s is the spacing of the hoops, as shown in Figure 3.6.

3.2.4 Fédération internationale du beton model code 2010

The fib (Fédération internationale du beton—International Federation for Structural Concrete) Model Code for Concrete Structures 2010 is an international code. The earlier version, CEB (European Committee for Concrete) Model Code 1990, served as an important basis for EC2.

The characteristic strength and strain of confined concrete can be determined from

$$\frac{f_{ck,c}}{f_{ck}} = 1.0 + 3.5\left(\frac{\sigma_2}{f_{ck}}\right)^{0.75} \tag{3.24}$$

$$\varepsilon_{c,c} = \varepsilon_c \left[1 + 5\left(\frac{f_{ck,c}}{f_{ck}} - 1\right)\right] \tag{3.25}$$

$$\varepsilon_{cu,c} = \varepsilon_{cu} + 0.2\sigma_2/f_{ck} \tag{3.26}$$

The effective lateral compressive stress can be determined by

$$\sigma_2 = \alpha\omega_c f_{cd} \tag{3.27}$$

with

$$\omega_c = \min\left\{\omega_y = \frac{A_{sy}f_{yhd}}{b_c s f_{cd}}, \omega_z = \frac{A_{sz}f_{yhd}}{h_c s f_{cd}}\right\} \tag{3.28}$$

$$\alpha = \left(1 - \frac{\sum b_i^2/6}{b_c h_c}\right)\left(1 - \frac{s}{b_c}\right)\left(1 - \frac{s}{h_c}\right) \tag{3.29}$$

where h_c and b_c are the depth and width of the confined core, respectively.

3.3 CONFINEMENT MODEL FOR HIGHLY CONFINED CONCRETE

In an SRC column, the steel section can provide lateral confining pressure through the contribution of bending stiffness of the steel flanges. The Mander model, Legeron model, fib model, or EC2 model, which are proposed for reinforced concrete, are applicable for HCC in SRC columns once the lateral confining stress from the steel section is clear. In SRC columns, HCC is confined by the steel section and the hoops together. Hence, the effective confining stress for HCC is obtained from:

$$f_{lh}' = f_{lp}' + f_{ls}' \tag{3.30}$$

where f_{lp}' and f_{ls}' are the effective lateral confining stress from the hoops and steel section, respectively.

3.3.1 Lateral confining stress from steel section

Concrete between the flanges of the structural steel section in SRC columns is more highly confined than concrete elsewhere. The effective confined zone is considered to be the total concrete area minus the area of the two parabolas with heights of one-quarter their length, as shown in Figure 3.7. El-Tawil (1996) suggested that the confinement effectiveness can be obtained from the following expression:

$$k'_e = \frac{\text{total concrete area between flanges-area of parabolas}}{\text{total concrete area between flanges}} \quad (3.31)$$

The confining pressure can be evaluated based on the resistance from the flanges. The moment applied at the base of the steel flange due to the expanding concrete should be equal to the resistance developed by the flanges, as shown in Figure 3.8. So

$$\frac{f'_{ly}}{k'_e} \frac{l^2}{2} = \frac{t^2 f_y}{6} \quad (3.32)$$

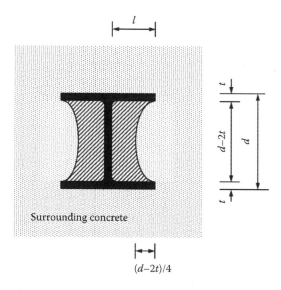

Figure 3.7 Highly confined region between steel flanges.

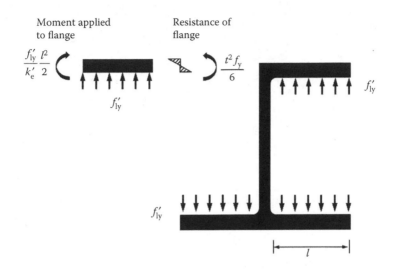

Figure 3.8 Confining pressure due to flanges.

The lateral confining pressure is given by

$$f'_{ly} = k'_e \frac{t^2 f_y}{3l^2} \tag{3.33}$$

where f'_{ly} is the additional confining pressure due to the steel section; f_y is the yield strength of structural steel; t is the thickness of the steel flange; and l is the length of the steel flange measured from the face of the web.

This lateral confining pressure is then added to the confining pressure previously calculated due to the lateral hoops. Thus, based on the total confining pressure, the strength and strain of highly confined concrete are determined from the confinement models for PCC.

3.4 MODIFIED CONFINEMENT MODEL FOR STEEL-REINFORCED CONCRETE COLUMNS

Modified confinement models for SRC columns were defined by combining the confinement model for PCC and the confinement effect from the steel section. The modified confinement model 1 was proposed by combining the Legeron model for PCC and the El-Tawil model for

HCC. According to the modified confinement model, the strength and strain of PCC and HCC are expressed as follows:

$$\left\{ \begin{aligned} f_{c,p} &= K_{p,I} f_{c,u} = \left(1 + 2.4 \left(\frac{f_{l,p}}{f_{c,u}} \right)^{0.7} \right) f_{c,u} \\ \varepsilon_{c,p} &= K_{p,II} \varepsilon_c = \left(1 + 35 \left(\frac{f_{l,p}}{f_{c,u}} \right)^{1.2} \right) \varepsilon_c \end{aligned} \right. \tag{3.34}$$

$$\left\{ \begin{aligned} f_{c,h} &= K_{h,I} f_{c,u} = \left(1 + 2.4 \left(\frac{f_{l,h}}{f_{c,u}} \right)^{0.7} \right) f_{c,u} \\ \varepsilon_{c,h} &= K_{h,II} \varepsilon_c = \left(1 + 35 \left(\frac{f_{l,h}}{f_{c,u}} \right)^{1.2} \right) \varepsilon_c \end{aligned} \right. \tag{3.35}$$

where $K_{p,I}$, $K_{p,II}$, $K_{h,I}$, and $K_{h,II}$ are confinement factors for PCC and HCC, respectively; $f_{l,p}$ and $f_{l,h}$ are the lateral confining stresses for PCC and HCC, respectively; and $f_{c,u}$ and ε_c are the strength and strain of unconfined concrete.

3.4.1 Effective lateral confining pressure on partially confined concrete

The effective lateral confining pressure for PCC from the hoops can be determined by the following:

$$f_{l,p} = k_e \rho_{sh} f_{r,h} \tag{3.36}$$

where

ρ_{sh} is the effective volume ratio of the hoops
k_e is the area ratio of the effectively confined area in PCC

$$k_e = \frac{\left(1 - \sum_{i=1}^{n} \frac{(b_i)^2}{6 b_c h_c} \right) \left(1 - \frac{s}{2 b_c} \right) \left(1 - \frac{s}{2 h_c} \right)}{1 - \rho_s} \tag{3.37}$$

where $f_{r,h}$ is the real stress of the hoops. Experimental studies conducted by Sheikh and Uzumeri (1980) showed that high-strength hoops may not yield when the confinement effect is insufficient. Thus, the real stress of the hoops needs to be determined, and the procedure proposed by Legeron and Paultre (2003) was adopted.

$$f_{r,h} = \begin{cases} f_{yh} & \kappa \leq 10 \\ \dfrac{0.25 f_{c,u}}{k_e \rho_{sh}(\kappa - 10)} \geq 0.43 \varepsilon_c E_s & \kappa \geq 10 \end{cases} \qquad (3.38)$$

where

$$\kappa = \frac{f_{c,u}}{k_e \rho_{sh} E_s \varepsilon_c} \qquad (3.39)$$

3.4.2 Effective lateral confining stress on highly confined concrete

The effective lateral confining stress for HCC can be determined by the following:

$$f_{l,h} = f_{l,p} + f_{l,s} \qquad (3.40)$$

where
$f_{l,s}$ is the effective lateral confining stress from the steel section

$$f_{l,s} = k'_e k_a f_{r,y} \qquad (3.41)$$

k'_e is the confinement effectiveness

$$k'_e = \frac{A_{c,f} - A_{c,r}}{A_{c,f}} \qquad (3.42)$$

$A_{c,f}$ is the total concrete area between the steel flanges
$A_{c,r}$ is the area of the two parabolas with heights of one-quarter their length
k_a is the confining factor

$$k_{\mathrm{a}} = \frac{t_f^2}{3l^2} \qquad\qquad\qquad (3.43)$$

where t_f is the steel flange thickness; l is the length of the steel flange measured from the face of the web; and $f_{r,y}$ is the real stress of the steel section.

The accuracy of the modified confinement models was verified by SRC column tests collected from the literature. The modified Legeron model can accurately predict the confinement effect for SRC columns based on the research work.

The resistance of SRC columns was determined by the full plastic method in accordance with EC4. Based on earlier discussion, the full plastic method overestimated the resistance of SRC columns with high-strength steel. Strain compatibility should be considered in the design. The peak strain and strength of confined concrete may be calculated by the modified confinement model. To ensure the yield strain of steel is less than the compressive strain of concrete, the maximum steel strength can be determined accordingly. Therefore, the resistance of SRC column may be predicted by the real stress of steel instead of the yield strength of steel, considering the confinement effect from the lateral hoops and the steel section and the strain compatibility.

Chapter 4

Concrete creep and shrinkage model

4.1 GENERAL

The ductility and strength of concrete can be improved by lateral confining pressure from the hoops and steel sections in SRC columns. Thus, with the increased compressive strain of confined concrete, the strain compatibility issue between high-strength steel and concrete, which was described in a previous chapter, can be solved. In addition to the confinement effect, the Germans are taking advantage of the time-dependent effects of creep and shrinkage to increase concrete compressive strain (Falkner et al.). This confirms that the yield strength of high-strength reinforcing steel can be fully exploited in RC columns if stress redistribution due to shrinkage and creep is taken into account.

The stress–strain curves of concrete that take into consideration the time-dependent effects for various loading rates and fatigue loading are shown in Figure 4.1. The main time-dependent effects are creep and shrinkage. There are several different calculation or prediction methods for time-dependent deformations available. In this book, the calculation method according to fib Model Code 2010 is used. Other prediction methods given in EC2 are also described.

4.2 CREEP AND SHRINKAGE

4.2.1 fib Model Code 2010

The CEB-FIP Model Code Committee published CEB Model Code 90, which provides procedures for estimating creep and shrinkage strains. Recently, the fib Model Code Committee published fib Model Code 2010, which contains some modifications of the procedures for estimating creep and shrinkage strains. The procedure will be presented here.

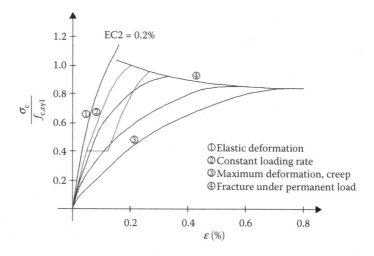

Figure 4.1 Time-dependent stress–strain curves for concrete.

The total strain of concrete $\varepsilon_c(t)$ at time t is:

$$\varepsilon_c(t) = \varepsilon_{ci}(t_0) + \varepsilon_{cc}(t) + \varepsilon_{cs}(t) + \varepsilon_{cT}(t) \tag{4.1}$$

where
$\varepsilon_{ci}(t_0)$ is the initial strain at loading;

$$\varepsilon_{ci}(t_0) = \frac{\sigma_c(t_0)}{E_{ci}(t_0)}$$

$\varepsilon_{cc}(t)$ is the creep strain at time t;
$\varepsilon_{cs}(t)$ is the shrinkage strain;
$\varepsilon_{cT}(t)$ is the thermal strain;
$E_{ci}(t_0)$ is the tangent modulus of elasticity at time t_0; and
$\sigma_c(t_0)$ is a constant stress at time t_0.

The initial strain is dependent on the age of the concrete at the time of loading and the initial stress. It can be time dependent if the load will change over time. But usually, this is considered constant.

4.2.1.1 Creep

In reality, creep is a nonlinear phenomenon. The creep coefficient can be used to describe the creep generated by a given duration of loading. The creep coefficient is defined as the ratio of the creep strain to the elastic strain of concrete at an age of 28 days under the same stress.

For a constant stress $\sigma_c(t_0)$, the creep strain is:

$$\varepsilon_{cc}(t,t_0) = \frac{\sigma_c(t_0)}{E_{ci}} \varphi(t,t_0) \tag{4.2}$$

where

$\varphi(t,t_0)$ is the creep coefficient; and

E_{ci} is the tangent modulus of elasticity at the age of 28 days.

The formula to determine the creep coefficient is empirical. It was calibrated based on laboratory concrete tests. Total creep is divided into basic creep and drying creep, reflecting the different associated physical mechanisms.

In this model, parameters that are normally known to the designer are taken into account, such as compressive strength of concrete, dimensions, mean relative humidity, age at loading, duration of loading, and cement type. Concrete creep depends on its composition and degree of hydration. Concrete creep decreases with decreasing water/cement ratio, decreasing cement paste content, increasing stiffness of the aggregates, and increasing degree of hydration at the age of loading.

The creep coefficient $\varphi(t,t_0)$ may be calculated from:

$$\varphi(t,t_0) = \varphi_{bc}(t,t_0) + \varphi_{dc}(t,t_0) \tag{4.3}$$

where

$\varphi_{bc}(t,t_0)$ is the basic creep coefficient;

$\varphi_{dc}(t,t_0)$ is the drying creep coefficient;

t is the age of concrete in days at the moment considered; and

t_0 is the age of concrete at loading in days adjusted.

The basic creep coefficient is

$$\varphi_{bc}(t,t_0) = \beta_{bc}(f_{cm}) \cdot \beta_{bc}(t,t_0) \tag{4.4}$$

with

$$\beta_{bc}(f_{cm}) = \frac{1.8}{(f_{cm})^{0.7}} \tag{4.5}$$

$$\beta_{bc}(t,t_0) = \ln\left(\left[\frac{30}{t_{0,adj}} + 0.035\right]^2 \cdot (t-t_0) + 1\right)$$ (4.6)

where

f_{cm} is the mean compressive strength at an age of 28 days; and $t_{0,adj}$ is the adjusted age at loading in days.

The drying creep coefficient is

$$\varphi_{dc}(t,t_0) = \beta_{dc}(f_{cm}) \cdot \beta(RH) \cdot \beta_{dc}(t_0) \cdot \beta_{dc}(t,t_0)$$ (4.7)

with

$$\beta_{dc}(f_{cm}) = \frac{412}{(f_{cm})^{1.4}}$$ (4.8)

$$\beta(RH) = \frac{1-(RH/100)}{\sqrt[3]{0.1 \cdot (h/100)}}$$ (4.9)

$$\beta_{dc}(t_0) = \frac{1}{0.1 + t_{0,adj}^{0.2}}$$ (4.10)

$$\beta_{dc}(t,t_0) = \left[\frac{(t-t_0)}{\beta_h + (t-t_0)}\right]^{\gamma(t_0)}$$ (4.11)

with

$$\gamma(t_0) = \frac{1}{2.3 + \left(3.5/\sqrt{t_{0,adj}}\right)}$$ (4.12)

$$\beta_h = 1.5h + 250\alpha_{f_{cm}} \le 1500\alpha_{f_{cm}}$$ (4.13)

$$\alpha_{f_{cm}} = \left(\frac{35}{f_{cm}}\right)^{0.5}$$ (4.14)

where

RH is the relative humidity of the ambient environment in %;

h is the notional size of the member;

$$h = \frac{2A_c}{u}$$

A_c is the area of the cross-section; and

u is the perimeter of the member.

The effect of cement type on the creep coefficient is taken into account by:

$$t_{0,\mathrm{adj}} = t_{0,\mathrm{T}} \cdot \left[\frac{9}{2 + t_{0,\mathrm{T}}^{1.2}} + 1 \right]^{\alpha} \geq 0.5 \qquad (4.15)$$

where

$t_{0,T}$ is the modified age of concrete at loading in days, which is equal to t_0 for a concrete temperature of $T = 20°C$; and

α is a coefficient that depends on the type of cement:

$\alpha = -1$ for strength class 32.5 N

$\alpha = 0$ for strength classes 32.5 R, 42.5 N

$\alpha = 1$ for strength classes 42.5 R, 52.5 N, 52.5 R

It should be noted that this equation was developed based on experimental results, primarily with CEM I and CEM III cements. If other cement types are used, this effect should be determined experimentally.

For stress in the range $0.4 f_{cm} (t_0) - 0.6 f_{cm} (t_0)$, the creep coefficient may be calculated by:

$$\varphi_\sigma(t,t_0) = \varphi(t,t_0) \cdot \exp[1.5(k_\sigma - 0.4)] \quad \text{for } 0.4 < k_\sigma < 0.6 \qquad (4.16)$$

where

$\varphi_\sigma(t,t_0)$ is the creep coefficient;

$\varphi(t,t_0)$ is the creep coefficient according to Equation 4.3; and

k_σ is the stress-strength ratio

$$k_\sigma = |\sigma_c|/f_{cm}(t_0)$$

The nonlinear behavior of concrete under high stresses mainly results from microcracking.

The creep coefficient is calculated from:

$$\varphi_t(t,t_0) = \varphi_0 \beta_c(t,t_0) \tag{4.27}$$

with

$$\varphi_0 = \varphi_{RH}\beta(f_{cm})\beta(t_0) \tag{4.28}$$

$$\beta(t,t_0) = \left[\frac{t-t_0}{(\beta_H + t - t_0)}\right]^{0.3} \tag{4.29}$$

where
φ_0 is the notional creep coefficient;

φ_{RH} is a factor to allow for the effect of relative humidity on the notional creep coefficient:

$$\varphi_{RH} = 1 + \frac{1 - RH/100}{0.1\sqrt[3]{h_0}} \qquad \text{for } f_{cm} \leq 35\,\text{MPa} \tag{4.30}$$

$$\varphi_{RH} = \left[1 + \frac{1 - RH/100}{0.1\sqrt[3]{h_0}}\alpha_1\right]\alpha_2 \qquad \text{for } f_{cm} > 35\,\text{MPa} \tag{4.31}$$

$\beta(f_{cm})$ is a factor to allow for the effect of concrete strength on the notional creep coefficient:

$$\beta(f_{cm}) = \frac{16.8}{\sqrt{f_{cm}}} \tag{4.32}$$

$\beta(t_0)$ is a factor to allow for the effect of concrete aging at loading:

$$\beta(t_0) = \frac{1}{\left(0.1 + t_0^{0.20}\right)} \tag{4.33}$$

h_0 is the notional size of the member in mm, where

$$h_0 = \frac{2A_c}{u}$$

u is the perimeter of the member in contact with atmosphere; and β_{H} is the coefficient depending on the relative humidity and the notional member size:

$$\beta_{\mathrm{H}} = 1.5[1 + (0.012RH)^{18}]h_0 + 250 \leq 1500 \qquad \text{for } f_{\mathrm{cm}} \leq 35 \quad (4.34)$$

$$\beta_{\mathrm{H}} = 1.5[1 + (0.012RH)^{18}]h_0 + 250\alpha_3 \leq 1500\alpha_3 \quad \text{for } f_{\mathrm{cm}} > 35 \quad (4.35)$$

$$\alpha_1 = \left(\frac{35}{f_{\mathrm{cm}}}\right)^{0.7} \quad \alpha_2 = \left(\frac{35}{f_{\mathrm{cm}}}\right)^{0.2} \quad \alpha_3 = \left(\frac{35}{f_{\mathrm{cm}}}\right)^{0.5} \qquad (4.36)$$

4.2.2.2 Shrinkage

In EC2, the total shrinkage strain is composed of the drying shrinkage strain and autogenous shrinkage strain. The drying shrinkage strain is a function of the migration of water through hardened concrete. The autogenous shrinkage strain develops during hardening of the concrete.
The total shrinkage strain can be calculated from:

$$\varepsilon_{\mathrm{cs}}(t) = \varepsilon_{\mathrm{cd}}(t) + \varepsilon_{\mathrm{ca}}(t) \qquad (4.37)$$

The drying shrinkage strain:

$$\varepsilon_{\mathrm{cd}}(t) = \beta_{\mathrm{ds}}(t, t_{\mathrm{s}}) \cdot k_{\mathrm{h}} \cdot \varepsilon_{\mathrm{cd},0} \qquad (4.38)$$

with

$$\beta_{\mathrm{ds}}(t, t_{\mathrm{s}}) = \frac{(t - t_{\mathrm{s}})}{(t - t_{\mathrm{s}}) + 0.04\sqrt{h_0^3}} \qquad (4.39)$$

$$\varepsilon_{\mathrm{cd}0} = 0.85 \left[(220 + 110\alpha_{\mathrm{ds}1}) \cdot \exp\left(-\alpha_{\mathrm{ds}2} \cdot \frac{f_{\mathrm{cm}}}{10}\right) \right] \cdot 10^{-6} \cdot \beta_{\mathrm{RH}} \qquad (4.40)$$

$$\beta_{\mathrm{RH}} = 1.55 \left[1 - \left(\frac{RH}{100}\right)^3 \right] \qquad (4.41)$$

Table 4.2 Values for k_h

h_0	k_h
100	1.0
200	0.85
300	0.75
≥ 500	0.70

where

k_h is a coefficient depending on the notional size h_0 (see Table 4.2);

α_{ds1} is coefficients dependent on the type of cement (see Table 4.1); and

α_{ds2} is coefficients dependent on the type of cement (see Table 4.1).

It should be noted that for cement Class R, the coefficient is taken as 0.11.

The autogenous shrinkage strain is given by:

$$\varepsilon_{ca}(t) = \varepsilon_{ca}(\infty) \cdot \beta_{as}(t) \tag{4.42}$$

with

$$\varepsilon_{ca}(\infty) = 2.5(f_{ck} - 10) \cdot 10^{-6} \tag{4.43}$$

$$\beta_{as}(t) = 1 - \exp(-0.2t^{0.5}) \tag{4.44}$$

4.3 LOAD REDISTRIBUTION

The calculations given in Section 4.2 are for unrestrained concrete. Usually, reinforcement is embedded in structural concrete and hence shortening is restrained. Load will be redistributed from concrete into the reinforcements because of strains caused by creep and shrinkage.

The stress in the reinforcements after redistribution can be calculated using the Trost method given by Ruesch (1983).

$$\sigma_s(t) = \sigma_{s0}\gamma_{sus} + \varepsilon_{cs} \cdot E_s \cdot \gamma_s(1 - \alpha) \tag{4.45}$$

with

$$\gamma_{\text{sus}} = 1 + \frac{(1-\alpha)\varphi}{1+\alpha\mu\varphi} \tag{4.46}$$

$$\gamma_{\text{s}} = \frac{1}{1+\alpha\mu\varphi} \tag{4.47}$$

where
α is the ratio of stiffness:

$$\alpha = n\frac{A_{\text{s}}}{A_{\text{tr}}} \tag{4.48}$$

A_{tr} is the transformed area:

$$A_{\text{tr}} = A_{\text{c}} + nA_{\text{s}} \tag{4.49}$$

n is the ratio of the modulus of elasticity:

$$n = \frac{E_{\text{s}}}{E_{\text{cm}}} \tag{4.50}$$

ρ is the reinforcement ratio:

$$\rho = \frac{A_{\text{s}}}{A_{\text{c}}} \tag{4.51}$$

ψ is the creep coefficient according to Section 4.2;
μ is the relaxation coefficient, $= 0.8$; and
σ_{s0} is the sustained stress.

The stress in concrete at the instant of loading can be determined by means of the loading and transformed area. Thus, the stress in the steel σ_{s0} is n times the stress in the concrete.

The redistributed stress is:

$$\Delta\sigma_{\text{s}} = \sigma_{\text{s}}(t) - \sigma_{\text{s0}} \tag{4.52}$$

And the additional strain in the reinforcement due to long-term effects is:

$$\Delta\varepsilon_s = \frac{\Delta\sigma_s}{E_s} = \frac{\sigma_s(t) - \sigma_{s0}}{E_s}$$

(4.53)

It should be noted that the reinforcement ratio influences the redistribution. With a low reinforcement ratio, the yield strength is quickly reached; thus, the redistribution load is low. With a high reinforcement ratio, the steel stress increases slowly, and the redistribution load increases (Falkner et al. 2008).

The time-dependent redistribution of the stresses in steel and concrete can be determined from Equation 4.52 or the more simplified method given by Wight and MacGregor (2012), as follows. At time t_0, the stress in the steel can be determined from:

$$\sigma_s(t_0) = n_0 \cdot \sigma_c(t_0)$$

(4.54)

where

A_{tr} is the transformed area, $A_{tr} = A_c + n_0 A_s$;
n_0 is the ratio of the modulus of elasticity, $n_0 = E_s/E_c(t_0)$; and
$\sigma_c(t_0)$ is the stress in the concrete at time t_0, $\sigma_c(t_0) = N/A_{tr}$.

At time t, an age-adjusted effective modulus is derived:

$$E_{caa}(t,t_0) = \frac{E_c(t_0)}{1 + \chi(t,t_0)(E_c(t_0)/E_c(28))\varphi(t,t_0)}$$

(4.55)

with

$$\chi(t,t_0) = \frac{t_0^{0.5}}{1 + t_0^{0.5}}$$

(4.56)

Thus, the stress in the steel can be determined from:

$$\sigma_s(t) = n_{aa} \cdot \sigma_c(t)$$

(4.57)

where

A_{traa} is the transformed area, $A_{traa} = A_c + n_{aa} A_s$;
n_{aa} is the ratio of the modulus of elasticity, $n_{aa} = E_s/E_{caa}(t,t_0)$; and
$\sigma_c(t)$ is the stress in the concrete, $\sigma_c(t) = N/A_{traa}$.

Therefore, the redistribution stress and additional strain due to long-term effects can be determined according to Equations 4.52 and 4.53. Compared to the method given by Ruesch, this method does not consider the time-dependent strain due to shrinkage. Therefore, this method is more conservative.

4.4 CONCRETE CREEP AND SHRINKAGE MODEL IN STEEL-REINFORCED CONCRETE COLUMNS

There is a strain-compatibility issue between high-strength reinforcing steel and concrete. Based on earlier research work, it is indicated that the yield strength of high-strength reinforcing steel can be fully exploited in RC columns if the redistribution due to shrinkage and creep is taken into account.

For SRC columns, the strain-compatibility issue also exists between high-strength steel and concrete. The issue can also be solved by taking advantage of time-dependent effects due to creep and shrinkage. However, tests on shrinkage and creep effects on the behavior of members are mostly done for RC members. Test data are insufficient to establish the validity of using the creep and shrinkage redistribution method for SRC columns. In order to further extend the application scope of the creep and shrinkage redistribution method to include SRC columns, more test data should be available, or more tests should be carried out. Therefore, it is recommended that the application of this method to SRC columns in real projects be done under a specialist's advice and direction.

Chapter 5

Design of steel-reinforced concrete columns

5.1 GENERAL

This book applies to SRC columns with steel grades up to S690 and concrete classes up to C90/105. The partial factors for calculating the design strength of concrete, structural steel, and reinforcing steel are shown in Table 5.1.

According to EN 1994-1-1, two methods are given for the design of composite columns: the general method and simplified method.

The general design method is applicable for composite members with a nonsymmetrical or nonuniform cross-section over the column length. Second-order effects should be considered in the general design method. As the materials used in composite columns follow different nonlinearity relationships, it is necessary to use numerical analysis for the design considerations (Structural & Conveyance Business, 2002). Generally, the amount of work is considerably large. Therefore, the general design method is not preferred in practical design.

The simplified design method is applicable for composite columns of doubly symmetrical and uniform cross-section over the column length. This method is based on some assumptions about the geometrical configurations of composite columns. Additionally, it makes use of the European buckling curves for bare steel columns. According to EN 1994-1-1, the limits of applicability of this method are given in the following. When these limits are not satisfied, the general design method mentioned above should be used.

The steel contribution ratio δ should satisfy the following conditions:

$$0.2 \leq \delta \leq 0.9; \tag{5.1}$$

$$\delta = \frac{A_a f_{yd}}{N_{pl,Rd}} \tag{5.~}$$

Table 5.1　Partial factors of materials

Concrete	Structural steel	Reinforcing steel	Shear connector
$\gamma_c = 1.5$	$\gamma_a = 1.0$	$\gamma_s = 1.15$	$\gamma_v = 1.25$

If δ is less than 0.2, the composite column should be treated as reinforced concrete, which may be designed in accordance with EN1992-1-1. If δ is larger than 0.9, the composite column is designed as a bare steel section, neglecting the effect of concrete.

The relative slenderness ratio of the composite column $\bar{\lambda}$ is limited to 2.0. This value results from test data of composite columns.

For fully encased steel sections, the thickness of the concrete cover must satisfy the following limits:

$$40 \text{ mm} \leq c_y \leq 0.4b, 40 \text{ mm} \leq c_z \leq 0.3h \qquad (5.3)$$

Based on test data for composite columns, the thickness limit of the concrete cover of composite columns is obtained. The limits are given to ensure that the bending stiffness of the steel section makes a significant contribution to the stiffness of the composite column. Sometimes, thicker concrete covers can be used in composite columns. Additionally, the thickness of the concrete cover of the steel section is required for design resistance to fire.

The ratio of the depth to the width of the composite cross-section should comply with the following limits:

$$0.2 < h_c/b_c < 5.0 \qquad (5.4)$$

The limit is to prevent the steel sections being susceptible to lateral–torsional buckling.

The maximum amount of longitudinal reinforcement is 6% of the concrete area. However, more longitudinal reinforcement is needed to ensure sufficient fire resistance.

Only the simplified design method will be considered further in this book. The general method could be implemented by means of finite-element analysis for SRC columns with high-strength materials.

The SRC columns in this publication should be checked for:

1. Resistance to local buckling
2. Resistance of cross-section

3. Resistance of member
4. Resistance of shear
5. Introduction of loads

5.2 LOCAL BUCKLING

Before calculating the plastic resistance of the composite cross-section, it should be checked whether local buckling of the steel section would occur. Therefore, the influence of local buckling of the steel section on the resistance should be considered in the design of composite columns.

To prevent local buckling, the limits given in Table 5.2 must be satisfied. For fully encased sections, the effects of local buckling may be neglected when the concrete cover thickness is adequate. The concrete cover of the flange of the steel section should be neither less than 40 mm nor less than $b/6$.

Table 5.2 Maximum values of geometrical parameters for preventing local buckling

Cross-section	Requirements
Fully concrete encased section	
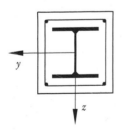	Minimum concrete cover $\geq$ (40 mm; $b/6$)
Partially encased I-sections	
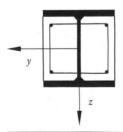	$\max\left(b/t_f\right) = 44\sqrt{\dfrac{235}{f_y}}$

5.3 AXIAL COMPRESSION

5.3.1 Resistance of cross-section

The design resistance to compression of an SRC column cross-section $N_{\text{pl,Rd}}$ is determined by adding the contributions of different components:

$$N_{\text{pl,Rd}} = A_a f_{yd} + 0.85 A_c \eta f_{cd} + A_s f_{sd} \tag{5.5}$$

where

A_a, A_c, and A_s are the cross-sectional areas of the steel, concrete, and reinforcing steel, respectively;

f_{yd}, f_{cd}, and f_{sd} are the design strengths of the steel, concrete, and reinforcing steel, respectively; and

η is the reduction factor for effective material strength.

For high-strength steel ($f_y > 460$ MPa), the design strength f_{yd} should be the real stress in steel instead of the yield strength. The real stress can be determined by multiplying the peak strain of the confined concrete by the modulus of elasticity.

For high-strength concrete with $f_{ck} > 50$ MPa, the effective compressive strength of concrete should be used in accordance with EC2. The effective strength is determined by multiplying the characteristic strength by a reduction factor η:

$$\eta = \begin{cases} 1.0 & f_{ck} \le 50\,\text{MPa} \\ 1.0 - (f_{ck} - 50)/200 & 50\,\text{MPa} < f_{ck} \le 90\,\text{MPa} \end{cases} \tag{5.6}$$

Accordingly, the secant modulus for HSC should be modified based on the effective strength:

$$E_{cm} = 22\left[(\eta f_{ck} + 8)/10\right]^{0.3} \tag{5.7}$$

The effective compressive strengths and modified moduli for various high-strength concrete classes are given in Tables 5.3 and 5.4.

5.3.2 Resistance of members

For SRC columns subjected to axial compression, the normal force N_{Ed} is required to satisfy the following expression:

$$\frac{N_{\text{Ed}}}{\chi N_{\text{pl,Rd}}} \le 1.0 \tag{5.8}$$

Table 5.3 Effective compressive strengths of HSC

	Concrete class				
	C55/67	C60/75	C70/85	C80/95	C90/105
Cylinder strength (MPa)	55	60	70	80	90
Reduction factor	0.975	0.95	0.9	0.85	0.8
Effective cylinder strength (MPa)	53.6	57	63	68	72

Table 5.4 Effective moduli of HSC

	Concrete class				
	C55/67	C60/75	C70/85	C80/95	C90/105
Modulus (GPa)	38.2	39.1	40.7	42.2	43.6
Reduction factor	0.99	0.99	0.97	0.96	0.94
Effective modulus (GPa)	38	38.6	39.6	40.4	41.1

The reduction factor χ for the relevant buckling mode in terms of relevant relative slenderness $\bar{\lambda}$ is given by

$$\chi = \frac{1}{\Phi + \sqrt{\Phi^2 - \bar{\lambda}^2}} \quad \text{but } \chi \leq 1.0 \tag{5.9}$$

with

$$\Phi = 0.5\left[1 + \alpha(\bar{\lambda} - 0.2) + \bar{\lambda}^2\right] \tag{5.10}$$

where

α is an imperfection parameter that is dependent on the buckling curve; and

$\bar{\lambda}$ is the relative slenderness.

The buckling curves and member imperfections for SRC columns are shown in Table 5.5. The imperfection factor α corresponding to the buckling curve can be determined from Table 5.6.

The relative slenderness $\bar{\lambda}$ is defined as:

$$\bar{\lambda} = \sqrt{\frac{N_{\text{pl,Rk}}}{N_{\text{cr}}}} \tag{5.1}$$

where

$N_{pl,Rk}$ is the characteristic value of the resistance of the SRC cross-section calculated by:

$$N_{pl,Rk} = A_a f_y + 0.85 A_c \eta f_{ck} + A_s f_{sk} \tag{5.12}$$

N_{cr} is the elastic critical buckling force for the relevant buckling mode:

$$N_{cr} = \frac{\pi^2 (EI)_{eff}}{L_{cr}^2} \tag{5.13}$$

L_{cr} is the buckling length of SRC columns for the relevant buckling mode. There is no guidance for the buckling length in Eurocode. Therefore, the buckling length of the SRC column can be taken as the member length L. Alternatively, the buckling lengths given in BS 5950: Part 1 are recommended. The buckling lengths and boundary conditions are as shown in Table 5.7 and Figure 5.1.

Table 5.5 Buckling curves and member imperfections

Cross-section	Axis of buckling	Buckling curve	Member imperfection
	y-y	b	L/200
	z-z	c	L/150
	y-y	b	L/200
	z-z	c	L/150

Table 5.6 Imperfection for buckling curves

Buckling curve	a	b	c
Imperfection factor α	0.21	0.34	0.49

Table 5.7 Buckling lengths for SRC columns

Nonsway mode		
End restraint in the plane under consideration by other parts of the structure		L_{cr}
Effectively held in position at both ends	(1) Effectively restrained in direction at both ends	0.7L
	(2) Partially restrained in direction at both ends	0.85L
	(3) Restrained in direction at one end	0.85L
	(4) Not restrained in direction at either end	1.0L

Sway mode		
One end	Other end	L_{cr}
Effectively held in position and restrained in direction	(5) Effectively restrained in direction	1.2L
	(6) Partially restrained in direction	1.5L
Not held in position	(7) Not restrained in direction	2.0L

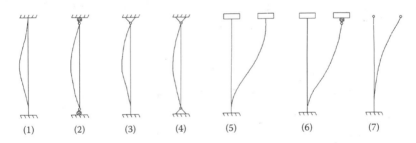

Figure 5.1 Buckling lengths for SRC columns.

For short-term loading, the effective elastic flexural stiffness $(EI)_{eff}$ of the composite cross-section about a principal axis (y or z) is obtained by adding up the flexural stiffness of the structural steel, the reinforcement, and the concrete of the cross-section:

$$(EI)_{eff} = E_a I_a + E_s I_s + K_e E_{cm} I_c \tag{5.14}$$

where

K_e is a correction factor, taken as 0.6;

I_a, I_c, and I_s are, respectively, the second moments of area of the structural steel section, the uncracked concrete section, and the reinforcement for the bending plane being considered;

E_{cm} is the elastic secant modulus of the concrete for short-term loading; for high-strength concrete, a reduced E_{cm} should be used.

For long-term loading, the creep and shrinkage of concrete will reduce the effective elastic flexural stiffness of the composite column. Therefore, the elastic critical normal force will be reduced. The effective modulus $E_{c,eff}$ depends on the age of the concrete at loading and the proportion of the design permanent load to the total axial load.

For long-term loading, the modulus of elasticity of concrete E_{cm} is reduced to the value $E_{c,eff}$ given by the following expression:

$$E_{c,eff} = E_{cm} \frac{1}{1 + (N_{G,Ed}/N_{Ed})\varphi_t} \tag{5.15}$$

where

$N_{G,Ed}$ is the part of this normal force that is permanent;

N_{Ed} is the total design normal force;

φ_t is the creep coefficient defined in EN1992-1-1, which depends on the age of the concrete at loading and at the time considered; see Section 4.2.

5.4 COMBINED COMPRESSION AND BENDING

5.4.1 Resistance of cross-section

According to EN 1994-1-1, the resistance of the composite cross-section and the corresponding interaction curve is calculated assuming rectangular stress blocks taking the design shear force into account, as shown in Figure 5.2. The tensile strength of the concrete is neglected. For a column with only a steel section, a continuous reduction of the moment resistance with an increase in the axial load can be observed from the interaction curve. However, in composite columns, as the compressive axial load can prevent concrete from cracking and make the composite cross-section more effective in resisting bending moments, an increase in the moment resistance may exist when the value of the axial load is relatively low.

An interaction curve of the cross-section of the composite column can be determined by using several possible positions of the plastic neutral axis and calculating the resistance to compression and moment resistance from the corresponding stress distributions. For the simplified method

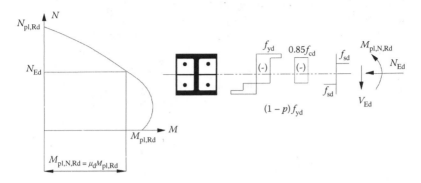

Figure 5.2 Interaction curve for combined compression and bending.

given in EN 1994-1-1, the interaction curve is replaced by a polygonal diagram, as shown in Figure 5.3. There are four possible positions of the plastic neutral axis in the interaction curve. In each position, the expressions for calculating the resistance to compression N_{Rd} and resistance to the bending moment about the relevant axis M_{Rd} can be determined.

The method to determine the point in the interaction curve is to assume a location for the plastic neutral axis. Then, the resistance N_{Rd} is calculated by summing the forces of individual components, and the moment resistance M_{Rd} is determined by taking the moments of the corresponding forces about the centroid of the uncracked section. Subsequently, the

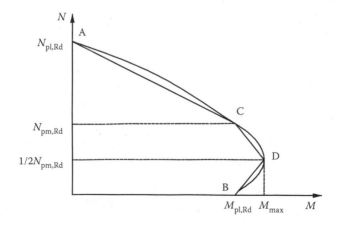

Figure 5.3 Simplified interaction curve of composite columns.

interaction curve is determined by moving the neutral axis from one point to another and finding N_{Rd} and M_{Rd} from the corresponding stress blocks.

As shown in Figure 5.3, point A represents pure compression and point B pure bending. The plastic neutral axis at point C is symmetrical to point B. The plastic neutral axis at point D is located in the center of the cross-section. The maximum moment resistance $M_{max,Rd}$ is achieved at point D. Figure 5.4 represents the stress distributions of the composite

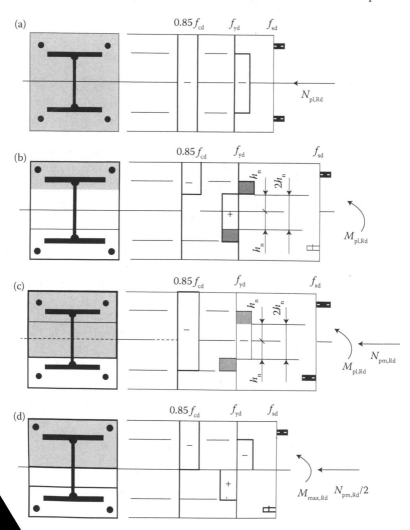

Figure 5.4 Stress distributions at different positions.

cross-section corresponding to the points A, B, C, and D. The significance of each of these points is described as follows:

1. *Point A—Pure compression*: Point A represents pure compression. The bending moment is zero at this point.

$$N_A = N_{pl,Rd}$$

$$M_A = 0$$

2. *Point B—Pure bending*: Point B represents the moment resistance of the composite section without compression force:

$$N_B = 0$$

$$M_B = M_{pl,Rd} = (W_{pa} - W_{pa,n})f_{yd} + (W_{ps} - W_{ps,n})f_{sd} \qquad (5.16)$$
$$+ 0.5(W_{pc} - W_{pc,n})\alpha_c f_{cd}$$

where W_{pa}, W_{ps}, and W_{pc} are the plastic section moduli for structural steel, reinforcement, and concrete, respectively; $W_{pa,n}$, $W_{ps,n}$, and $W_{pc,n}$ are the plastic section moduli for the structural steel, reinforcement, and concrete within region $2h_n$, respectively; and α_c is taken as 0.85.

3. *Point C*: The moment resistance is equal to point B. The axial compression force of the SRC column is obtained from:

$$N_C = N_{pm,Rd} = A_c \alpha_c f_{cd} \qquad (5.17)$$

$$M_C = M_B = M_{pl,Rd}$$

The expressions for calculating the corresponding resistance may be determined by combining the stress distributions at points B and C. The compression area of the concrete at point B and the tension area of the concrete at point C are identical. The moment resistance at point C and the moment resistance at point B are also identical. However, the resistance at point C is equal to the resistance to compression of the concrete alone, $N_{pm,Rd}$, which is different from that at point B.

4. *Point D*: The compression and moment resistance of the SRC column can be obtained from:

$$N_C = 0.5N_{pm,Rd}$$

$$M_D = M_{max,Rd} = W_{pa}f_{yd} + W_{ps}f_{sd} + 0.5W_{pc}\alpha_c f_{cd} \qquad (5.$$

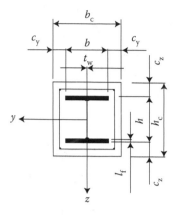

Figure 5.5 Typical SRC column section.

After the determination of the four points, the interaction curve can be obtained.

The values of the parameters mentioned in the above equations for the SRC column sections shown in Figure 5.5 are given in the following:

1. *Major axis*: The plastic section modulus of structural steel may be taken from tables or calculated from:

$$W_{\mathrm{pa}} = \frac{(h - 2t_{\mathrm{f}})^2 t_{\mathrm{w}}}{4} + bt_{\mathrm{f}}(h - t_{\mathrm{f}}) \tag{5.19}$$

The plastic section modulus of concrete is calculated from:

$$W_{\mathrm{pc}} = \frac{b_{\mathrm{c}}h_{\mathrm{c}}^2}{4} - W_{\mathrm{pa}} - W_{\mathrm{ps}} \tag{5.20}$$

The plastic section modulus of longitudinal reinforcement is determined by:

$$W_{\mathrm{ps}} = \sum_{i=1}^{n} A_{\mathrm{s,i}}[e_i] \tag{5.21}$$

where e_i are the distances of the reinforcement bars of the area $A_{\mathrm{s,i}}$ to the relevant middle line (y-axis or z-axis).

For different neutral axis positions, h_n and $W_{pa,n}$ are calculated from:

a. Neutral axis in the web, $h_n \leq h/2 - t_f$:

$$h_n = \frac{A_c \alpha_c f_{cd} - A_{sn}(2f_{sd} - \alpha_c f_{cd})}{2b_c \alpha_c f_{cd} + 2t_w(2f_{yd} - \alpha_c f_{cd})} \tag{5.22}$$

$$W_{pa,n} = t_w h_n^2 \tag{5.23}$$

where A_{sn} is the sum of the area of reinforcements within the region of depth $2h_n$.

b. Neutral axis in the flange, $h/2 - t_f < h_n < h/2$:

$$h_n = \frac{\begin{aligned}&A_c \alpha_c f_{cd} - A_{sn}(2f_{sd} - \alpha_c f_{cd}) \\ &+ (b - t_w)(h - 2t_f)(2f_{yd} - \alpha_c f_{cd})\end{aligned}}{2b_c \alpha_c f_{cd} + 2b(2f_{yd} - \alpha_c f_{cd})} \tag{5.24}$$

$$W_{pa,n} = bh_n^2 - \frac{(b - t_w)(h - 2t_f^2)}{4} \tag{5.25}$$

c. Neutral axis outside the steel section, $h/2 \leq h_n \leq h_c/2$:

$$h_n = \frac{A_c \alpha_c f_{cd} - A_{sn}(2f_{sd} - \alpha_c f_{cd}) - A_a(2f_{yd} - \alpha_c f_{cd})}{2b_c \alpha_c f_{cd}} \tag{5.26}$$

$$W_{pa,n} = W_{pa} \tag{5.27}$$

The plastic modulus of concrete within the region $2h_n$ is given by:

$$W_{pcn} = b_c h_n^2 - W_{pa,n} - W_{ps,n} \tag{5.28}$$

The plastic modulus of the total reinforcement within the region $2h_n$ is given by:

$$W_{ps,n} = \sum_{i=1}^{n} A_{sn,i}[e_{z,i}] \tag{5.29}$$

where $A_{sn,i}$ is the longitudinal reinforcement area within the region $2h_n$; and $e_{z,i}$ is the distance from the middle line of the major axis.

2. *Minor axis*: The plastic section modulus of the structural steel may be taken from tables or calculated from:

$$W_{pa} = \frac{(h - 2t_f)t_w^2}{4} + \frac{2b^2 t_f}{4} \tag{5.}$$

The plastic section modulus of concrete may be calculated from:

$$W_{pc} = \frac{b_c^2 h_c}{4} - W_{pa} - W_{ps} \tag{5.31}$$

The plastic section modulus of longitudinal reinforcements is calculated from:

$$W_{ps} = \sum_{i=1}^{n} A_{s,i}[e_i] \tag{5.32}$$

where e_i are the distances of the reinforcement bars of the area $A_{s,i}$ to the relevant middle line (y-axis or z-axis).

For different neutral axis positions, h_n and $W_{pa,n}$ are calculated from:

a. Neutral axis in the web, $h_n \leq t_w/2$:

$$h_n = \frac{A_c \alpha_c f_{cd} - A_{sn}(2f_{sd} - \alpha_c f_{cd})}{2h_c \alpha_c f_{cd} + 2h(2f_{yd} - \alpha_c f_{cd})} \tag{5.33}$$

$$W_{pa,n} = h h_n^2 \tag{5.34}$$

b. Neutral axis in the flange, $t_w/2 < h_n < b/2$:

$$h_n = \frac{A_c \alpha_c f_{cd} - A_{sn}(2f_{sd} - \alpha_c f_{cd}) + t_w(2t_f - h)(2f_{yd} - \alpha_c f_{cd})}{2h_c \alpha_c f_{cd} + 4t_f(2f_{yd} - \alpha_c f_{cd})} \tag{5.35}$$

$$W_{pa,n} = 2t_f h_n^2 - \frac{t_w^2(h - 2t_f)}{4} \tag{5.36}$$

c. Neutral axis outside the steel section, $b/2 \leq h_n \leq b_c/2$:

$$h_n = \frac{A_c \alpha_c f_{cd} - A_{sn}(2f_{sd} - \alpha_c f_{cd}) - A_a(2f_{yd} - \alpha_c f_{cd})}{2h_c \alpha_c f_{cd}} \tag{5.37}$$

$$W_{pa,n} = W_{pa} \tag{5.38}$$

The plastic modulus of concrete within the region $2h_n$ is given by:

$$W_{pc,n} = h_c h_n^2 - W_{pa,n} - W_{ps,n} \tag{5.39}$$

The plastic modulus of the total reinforcement within the region $2h_n$ is given by:

$$W_{ps,n} = \sum_{i=1}^{n} A_{sn,i}[e_{y,i}] \qquad (5.40)$$

5.4.2 Second-order effects and member imperfections

In the design of composite columns, it is necessary to ascertain whether the second-order effects need to be taken into account. According to EN 1994-1-1, the second-order effects on bending moments for composite columns may be neglected if the following conditions are satisfied:

$$\alpha_{cr} = \frac{N_{cr,eff}}{N_{Ed}} \geq 10 \qquad (5.41)$$

where the elastic critical force $N_{cr,eff}$ for the relevant axis and corresponding to the effective stiffness is found from the following equations.

$$N_{cr,eff} = \frac{\pi^2 (EI)_{eff,II}}{L^2} \qquad (5.42)$$

The effective flexural stiffness takes account of the long-term effects and should be determined from the following expression:

$$(EI)_{eff,II} = K_o(E_a I_a + E_s I_s + K_{e,II} E_{cm} I_c) \qquad (5.43)$$

where
K_o is a calibration factor that should be taken as 0.9; and
$K_{e,II}$ is a correction factor that should be taken as 0.5.

If the second-order effects are neglected, the design bending moment M_{Ed} for the composite column is the maximum value given by first-order member analysis. It does not mean that the influence of the member imperfection on the bending resistance can be neglected, though the second-order effect is ignored.

The member imperfections given in Table 5.5 are related to the length L of the composite column. However, the distribution of the bending moment along the length of the composite column does not affect the values of imperfection. For the initial imperfection e_0 caused by the design axial load N_{Ed} on a composite column, the bending moment will be $N_{Ed}e_0$.

Generally, the second-order effects on the bending moment will usually need to be considered in the design because most composite columns are relatively slender. Second-order effects may be considered by multiplying the greatest first-order design bending moment M_{Ed} by a factor k, given by:

$$k = \frac{\beta}{1 - N_{Ed}/N_{cr,eff}}, \quad \geq 1.0 \tag{5.44}$$

where β is an equivalent moment factor, given in Table 5.8.

The equivalent moment factor β, as shown in Table 5.8, is related to the shape of the bending moment diagram. In the design of composite columns, the value of the factor β should be at least 0.44, which is used to ensure sufficient protection against snap-through buckling.

In the process of determining the maximum bending moment of the composite column, the moments caused by second-order effects and imperfections are found separately and can be added together. Figure 5.6a and b show the moments from first-order analysis and initial member

Table 5.8 Factors for the determination of moments to second-order theory

Moment distribution	Moment factors β	Comment
M_E (two curved distributions)	First-order bending moments from member imperfection or lateral load: $\beta = 1.0$	M_{Ed} is the maximum bending moment within the column length ignoring second-order effects
M_E $r\,M_{Ed}$ $-1 \leq r \leq 1$	End moments: $\beta = 0.66 + 0.44r$ but $\beta \geq 0.44$	M_{Ed} and $r\,M_{Ed}$ are the end moments from first-order or second-order global analysis

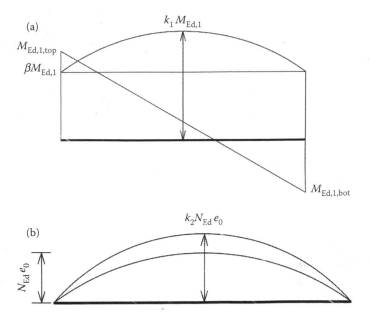

Figure 5.6 Moments from first-order analysis and member imperfection. (a) First order moment, (b) moment from imperfection.

imperfection, respectively. As there are two moment distributions to be considered, two corresponding factors (k_1 and k_2) are used to determine the maximum bending moment. The design bending moment for the composite column length considering both second-order effects and imperfection is given by:

$$M_{Ed,max} = k_1 M_{Ed,1} + k_2 N_{Ed} e_0 \qquad (5.45)$$

where

$M_{Ed,1}$ is the maximum first-order design moment in the column length; and

e_0 is the member imperfection given in Table 5.5.

According to EN 1994-1-1, a further provision for Equation 5.44 is that k must be greater than or equal to 1.0. The limit is given for ensuring that the design bending moment is the larger end moment M_{Ed}. For a single distribution of bending moments, it is applicable for the design of the column. However, for a combination of second-order effects and memb~ imperfections, it is overconservative to apply this limit, and k is us~

less than 1.0. Thus, the factor k_1 for determining the bending moment $k_1 M_{Ed}$ needs not to be increased to 1.0.

For the first-order bending moment from the member imperfection $N_{Ed} e_0$, the value of β equals 1.0, as shown in Table 5.8. From Equation 5.44, the factor k_2 always exceeds 1.0. Thus, the factor k_2 is usually different from k_1. The design bending moment can be determined by combining the moment $k_2 N_{Ed} e_0$ and $k_1 M_{Ed,1}$.

5.4.3 Resistance of members in combined compression and uniaxial bending

According to EN 1994-1-1, based on the interaction curve of the composite cross-section, the resistance of members in combined compression and uniaxial bending can be checked. With axial load N_{Ed}, a moment resistance $\mu_d M_{pl,Rd}$ is obtained from the interaction curve, as shown in Figure 5.7. However, it is nonconservative, and thereby the moment resistance is reduced by using a factor α_M that is dependent on the grade of the structural steel.

When the column is subjected to compression and uniaxial bending, the following equation based on the interaction curve should be satisfied:

$$\frac{M_{Ed}}{M_{pl,N,Rd}} = \frac{M_{Ed}}{\mu_d M_{pl,Rd}} \leq \alpha_M \qquad (5.46)$$

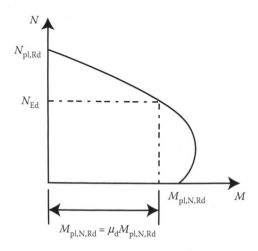

re 5.7 Interaction curve for compression and uniaxial bending.

where

M_{Ed} is the maximum bending moment, including imperfections and second-order effects;

$M_{\mathrm{pl,N,Rd}}$ is the plastic bending resistance, taking into account the normal force N_{Ed};

$M_{\mathrm{pl,Rd}}$ is the plastic bending resistance, given by point B in Figure 4.5;

α_{M} is a coefficient factor:

For steel grades S235 to S355, the coefficient α_{M} should be taken as 0.9;

For steel grades S420 to S690, it should be taken as 0.8; and

μ_{d} is a factor obtained from the interaction curve.

The value μ_{d} can be interpolated according to Figure 5.7.

1. For $N_{\mathrm{Ed}} \leq N_{\mathrm{pm,Rd}}/2$

$$\mu_{\mathrm{d}} = 1 + \frac{2N_{\mathrm{Ed}}}{N_{\mathrm{pm,Rd}}} \left(\frac{M_{\mathrm{max,Rd}}}{M_{\mathrm{pl,Rd}}} - 1 \right) \tag{5.47}$$

2. For $N_{\mathrm{pm,Rd}}/2 < N_{\mathrm{Ed}} \leq N_{\mathrm{pm,Rd}}$

$$\mu_{\mathrm{d}} = 1 + \frac{2(N_{\mathrm{pm,Rd}} - N_{\mathrm{Ed}})}{N_{\mathrm{pm,Rd}}} \left(\frac{M_{\mathrm{max,Rd}}}{M_{\mathrm{pl,Rd}}} - 1 \right) \tag{5.48}$$

3. For $N_{\mathrm{Ed}} > N_{\mathrm{pm,Rd}}$

$$\mu_{\mathrm{d}} = \frac{N_{\mathrm{pl,Rd}} - N_{\mathrm{Ed}}}{N_{\mathrm{pl,Rd}} - N_{\mathrm{pm,Rd}}} \tag{5.49}$$

It is obvious that values of μ_{d} obtained from the interaction curve may exceed 1.0 in the region around point D, as shown in Figure 5.3. In practice, values of μ_{d} greater than 1.0 should not be used, except that the bending moment M_{Ed} depends directly on the action of the normal force N_{Ed}, for example, where the moment M_{Ed} results from an eccentricity of the normal force N_{Ed}.

5.4.4 Resistance of members in combined compression and biaxial bending

For composite columns under combined compression and biaxial bendi
it should be checked first whether the resistance under uniaxial ber

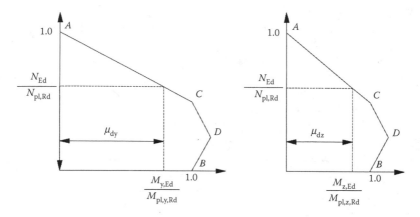

Figure 5.8 Interaction curves for combined compression and biaxial bending.

in the major and minor axes satisfies the requirement and its biaxial bending behavior. In the process of checking, it needs to be decided in which plane bending failure is expected to occur. The imperfection is considered for this plane only. For the other plane of bending, the influence of member imperfections on the bending moment is neglected.

For a column under combined compression and biaxial bending, the following equations based on the interaction curve in Figure 5.8 should be satisfied:

$$\frac{M_{y,Ed}}{\mu_{dy} M_{pl,y,Rd}} \le \alpha_{M,y}$$

$$\frac{M_{y,Ed}}{\mu_{dy} M_{pl,y,Rd}} \le \alpha_{M,y} \tag{5.50}$$

$$\frac{M_{z,Ed}}{\mu_{dz} M_{pl,z,Rd}} \le \alpha_{M,z} \tag{5.51}$$

$$\frac{M_{y,Ed}}{\mu_{dy} M_{pl,y,Rd}} + \frac{M_{z,Ed}}{\mu_{dz} M_{pl,z,Rd}} \le 1.0 \tag{5.52}$$

where

$M_{pl,y,Rd}$ and $M_{pl,z,Rd}$ are the plastic bending resistances of the relevant plane of bending;

$M_{y,Ed}$ and $M_{z,Ed}$ are the maximum bending moment including imperfections and second-order effects, if necessary;
μ_{dy} and μ_{dz} are the factors obtained from the interaction curve of the relevant plane of bending; and
$\alpha_{M,y}$, $\alpha_{M,z}$ $\alpha_{M,y} = \alpha_{M,z} = \alpha_M$.

5.5 RESISTANCE TO SHEAR FORCE

The shear force V_{Ed} may be distributed into $V_{a,Ed}$ acting on the structural steel section and $V_{c,Ed}$ acting on the reinforced concrete section, which are:

$$V_{a,Ed} = V_{Ed} \frac{M_{pl,a,Rd}}{M_{pl,Rd}} \tag{5.53}$$

$$V_{c,Ed} = V_{Ed} - V_{a,Ed} \tag{5.54}$$

where
$M_{pl,a,Rd}$ is the plastic resistance moment of the steel section; and
$M_{pl,Rd}$ is the plastic resistance moment of the composite section.

For simplicity, the design transverse shear force V_{Ed} is assumed to completely act on the steel section alone.

According to EN 1994-1-1, the influence of transverse shear forces on the resistance to the bending moment and normal force need to be considered when determining the interaction curve if the shear force $V_{a,Ed}$ on the steel section exceeds 50% of the design shear resistance $V_{pl,a,Rd}$ of the steel section. The influence should be determined by a reduced design steel strength $(1 - \rho)f_{yd}$ in the shear area A_v.

The reduction factor can be calculated from

$$\rho = \left(2\frac{V_{Ed}}{V_{Rd}} - 1 \right)^2 \tag{5.55}$$

The design shear resistance of steel is given by

$$V_{pl,a,Rd} = A_v \frac{f_{yd}}{\sqrt{3}} \tag{5.56}$$

where A_v is the shear area of the steel section.

However, no reduction in the web thickness is necessary if the shear force $V_{a,Ed}$ on the steel section doesn't exceed 50% of the design shear resistance $V_{pl,a,Rd}$.

5.6 INTRODUCTION OF LOAD

If loads are introduced into a composite column, it should be ensured that the steel section and concrete of the composite cross-section are loaded according to the corresponding resistance within the introduction length.

The load applied to the composite column can be simply distributed to the steel and concrete using the following expressions:

$$N_{c,Ed} = N_{Ed}(1 - \delta) \tag{5.57}$$

$$N_{a,Ed} = N_{Ed} - N_{c,Ed} \tag{5.58}$$

where

N_{Ed} is the design axial load;

$N_{c,Ed}$ is the design axial load applied to the concrete and reinforcement; and

$N_{a,Ed}$ is the design axial load applied to the steel section.

Figure 5.9 shows a typical beam–column connection and the introduction length l_v. According to EN 1994-1-1, the introduction length l_v should not exceed $2d$ or $L/3$, where d is the minimum transverse dimension of the column and L is the column length.

In regions of load introduction, if the shear stress τ_{Ed} exceeds the design shear strength τ_{Rd}, shear connection is required. No well-established method is given in EN 1994-1-1 for calculating longitudinal shear stress τ_{Ed} at the interface of steel and concrete. The longitudinal shear stress at the interface between steel and concrete can be estimated by elastic analysis of the uncracked composite section.

The shear stress τ_{Ed} is usually determined by the following expression:

$$\tau_{Ed} = N_{c,Ed}/p_a l_v \tag{5.59}$$

where

$N_{c,Ed}$ is the force that caused the shear at the interface of the steel and concrete section; and

P_a is the perimeter of the steel section at the interface of the steel and concrete section.

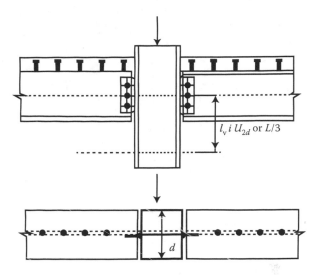

Figure 5.9 Force transfer in composite beam–column connection.

According to EN 1994-1-1, the design shear strength τ_{Rd} due to bond and friction is given in Table 5.9. For fully concrete-encased steel sections, where small steel I-sections are provided and the column is mainly concrete, higher values of τ_{Rd} may be used. Unless verified by tests, for a completely encased section, the value τ_{Rd} may be increased by multiplying by a factor β_c given by:

$$\beta_c = 1 + 0.02c_z(1 - c_{z,min}/c_z) \leq 2.5 \tag{5.60}$$

where
c_z is the nominal value of the concrete cover in mm; and
$c_{z,min}$ is the minimum concrete cover, 40 mm.

If τ_{Ed} is less than the design shear strength τ_{Rd}, it is not necessary to provide shear connectors between the steel and concrete. The transverse reinforcement can provide sufficient protection against local failure of the

Table 5.9 Design shear strength τ_{Rd}

Type of cross-section	τ_{Rd} (N/mm²)
Completely concrete-encased steel sections	0.30
Partially encased sections	0.20

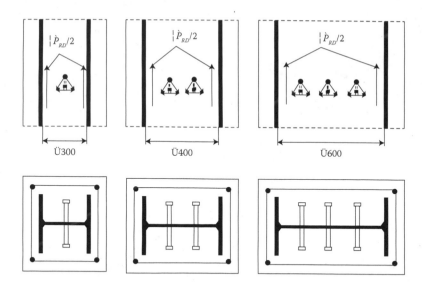

Figure 5.10 Additional frictional forces in composite columns by use of headed studs.

composite member. However, if the design shear strength τ_{Rd} is exceeded at the interface between the steel and concrete, shear connectors should be provided in the load introduction region.

For fully or partially concrete-encased steel sections, if the bond strength between the steel section and concrete is insufficient for transmitting the force, shear connectors attached to the web of the steel section should be provided. Additional resistance can be developed from the prevention of lateral expansion of concrete by the steel flange, which may be added to the resistance of the shear connectors. The additional resistance may be taken as $\mu P_{Rd}/2$ on each flange and each horizontal row of studs, as shown in Figure 5.10. The relative coefficient of friction μ may be taken as 0.50. Actually, the value of μ is dependent on the degree of confinement to which the concrete between the flanges of the section is subjected. In addition, the distance between the flanges should not exceed the values given in Figure 5.10.

5.7 SUMMARY

The current EC4 method can be safely extended to the design of SRC columns with steel strength up to 690 N/mm² and concrete compressive cylinder strength up to 90 N/mm^2, with the following modifications:

1. The strain-compatibility issue between steel and concrete materials must be observed. The strain-compatibility method may be used to select the steel grade and concrete class for the design of SRC columns to avoid crushing of the core concrete before the steel yields.
2. A strength reduction factor should be applied for high-strength concrete ($f_{ck} > 50$ MPa). In addition, the secant modulus of concrete should be modified accordingly.

Although this design guide may be applied to SRC columns with high-strength steel, more tests are needed to justify the use of the present method. A conservative approach to designing SRC columns with high-strength materials is to adopt the reduced concrete strength and real stress of steel considering the strain-compatibility issue.

A design flow chart is given in Appendix B for the design of SRC members with an extension of the EC4 Method to C90/105 Concrete and S690 Steel.

Chapter 6

Fire design

6.1 GENERAL

Composite members shall comply with criteria R, E, and I in accordance
with EN 1994-1-2. Three approaches are given in EN 1994-1-2 to assess
structural behavior in a fire design situation. The three levels of methods are:

1. *Tabular methods*: Presented as tabular data for specific types of
   structural members. Design tables are relatively restrictive but
   cover common cases of design.
2. *Simple calculation models*: This method is generally hand-
   calculation method. It is based on well-established principles, such
   as plastic analysis of the section. This method is used for general
   design and will lead to more economic design than the tabular
   methods.
3. *Advanced calculation models*: This method simulates the behavior
   of the global structure or isolated member. This method is
   appropriate only for computer analysis, not for general design.

The tabular method and the simple calculation method are used only for
specific types of members under prescribed situations. The temperature
distribution is assumed to be the same over the full length. Therefore,
these two methods are conservative.

6.2 TEMPERATURE–TIME CURVES IN FIRE

In a fire resistance test, the member is exposed to the standard
temperature–time curve according to EN 1991-1-2. The temperature–
time curve defined in ISO 834 or EN 1991-1-2 is:

$$\theta_g = 20 + 345 \log_{10}^{(8t+1)} \tag{6.1}$$

where

θ_g is the gas temperature in the fire compartment [°C]; and
t is the time in minutes.

The standard curve is characterized by atmosphere temperatures that rise continuously with time at a diminishing rate, as shown in Figure 6.1.

An alternative method to the use of the standard fire curve is to model a natural fire on the basis of a parametric temperature–time curve. The parametric temperature can be determined in accordance with EN 1991-1-2, which relates to compartment dimensions, opening area, fire load density, and thermal properties (density, specific heat, and thermal conductivity).

6.3 PERFORMANCE OF MATERIAL AT ELEVATED TEMPERATURE

Materials (concrete, structural steel, and reinforcing steel) lose strength at elevated temperatures. The thermal and mechanical properties of steel and concrete may be referred to EN 1994-1-2, EN 1992-1-2, and EN 1993-1-2.

6.3.1 Structural steel

The mechanical properties of structural steel are given in EN 1994-1-2 and EN 1993-1-2. EC3 and EC4 are consistent in terms of material properties.

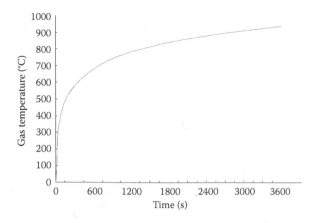

Figure 6.1 Temperature curve for ISO 834 standard fire.

Table 6.1 gives the reduction factors for the stress–strain relationship for steel at elevated temperatures. These reduction factors are: (1) reduction factor for the slope of the linear elastic range $k_{E,\theta}$, (2) reduction factor for the proportional limit $k_{p,\theta}$, (3) reduction factor for the effective yield strength $k_{y,\theta}$, and (4) reduction factor for the yield strength considering strain hardening $k_{u,\theta}$. The stress–strain relationships with temperature for the steel grade S235 are shown in Figure 6.2, with a maximum strain of 2%.

6.3.2 Concrete

The mechanical properties of concrete are given in EN 1994-1-2 and EN 1992-1-2. The properties of concrete at elevated temperatures are related to the aggregate type. Calcareous concrete has a lower thermal conductivity than concrete with siliceous aggregate. However, only the properties of concrete with siliceous aggregates are given in EN 1994-1-2. The properties of concrete with calcareous aggregates are referred to in EN 1992-1-2.

Table 6.2 gives the reduction factors for the stress–strain relationship for concrete at elevated temperatures. Concrete with siliceous aggregates is used in Table 6.2. The stress–strain relationships for concrete with siliceous aggregates are shown in Figure 6.3, with a maximum strain of 5%.

Table 6.1 Reduction factors for stress–strain relationships of structural steel at elevated temperatures

Temperature (°C)	$k_{E,\theta} = E_{a,\theta}/E_a$	$k_{p,\theta} = f_{ap,\theta}/f_{ay}$	$k_{y,\theta} = f_{ay,\theta}/f_{ay}$	$k_{u,\theta} = f_{au,\theta}/f_{ay}$
20	1.0	1.0	1.0	1.25
100	1.0	1.0	1.0	1.25
200	0.90	0.807	1.0	1.25
300	0.80	0.613	1.0	1.25
400	0.70	0.42	1.0	
500	0.60	0.36	0.78	
600	0.31	0.18	0.47	
700	0.13	0.075	0.23	
800	0.09	0.05	0.11	
900	0.0675	0.0375	0.06	
1000	0.045	0.025	0.04	
1100	0.0225	0.0125	0.02	
1200	0	0	0	

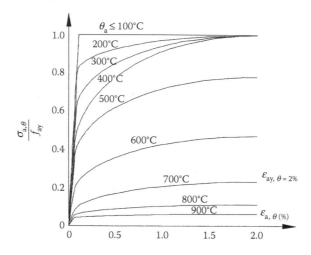

Figure 6.2 Stress–strain relationships with temperature for steel grade 235.

Table 6.2 Reduction factors for stress–strain relationships of
concrete at elevated temperatures

Temperature (°C)	The reduction factor for the compressive strength $k_{c,\theta} = f_{c,\theta}/f_c$	$\varepsilon_{cu,\theta}$
20	1.0	0.0025
100	1.0	0.004
200	0.95	0.0055
300	0.85	0.007
400	0.75	0.01
500	0.60	0.015
600	0.45	0.025
700	0.30	0.025
800	0.15	0.025
900	0.08	0.025
1000	0.04	0.025
1100	0.01	0.025
1200	0	–

Concrete has a lower thermal conductivity than steel, so it can provide relatively good insulation to steel sections and reinforcement in SRC columns. The fire resistance of SRC members is based on the strength reduction of steel, which depends on the concrete cover. However, concrete may suffer from the phenomenon of spalling at elevated temperatures,

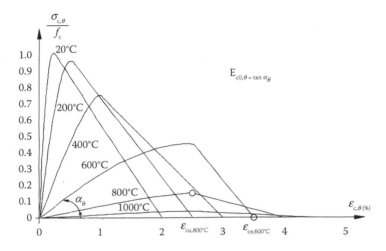

Figure 6.3 Stress–strain relationships with temperature for concrete with siliceous aggregates.

which is a breaking away of the concrete cover. This may lead to the exposure of steel and reinforcements to fire. Therefore, supplementary mesh or links are often required in order to prevent spalling of the concrete.

6.3.3 Reinforcing steel

The mechanical properties of reinforcing steel are given in EN 1994-1-2 and EN 1992-1-2. The strength and deformation properties of reinforcing steel at elevated temperatures may be obtained from the same model as structural steel. Table 6.3 gives the reduction factors for the stress–strain relationship for cold-worked reinforcing steel at elevated temperatures. The properties of hot-rolled reinforcing steel may be referred to in EN 1992-1-2.

6.4 LOAD LEVEL

The concept of load level is very important in fire design. According to EN 1994-1-2, the load levels are defined by the ratio between the relevant design effect of actions and the design of resistance in a fire situation:

$$\eta_{fi,t} = \frac{E_{fi,d,t}}{R_d} \tag{6.2}$$

Table 6.3 Reduction factors for stress–strain relationships of
cold-worked reinforcing steel at elevated temperatures

Temperature (°C)	$k_{E,\theta} = E_{s,\theta}/E_s$	$k_{p,\theta} = f_{sp,\theta}/f_{sy}$	$k_{y,\theta} = f_{sy,\theta}/f_{sy}$
20	1.0	1.0	1.0
100	1.0	0.96	1.0
200	0.87	0.92	1.0
300	0.72	0.81	1.0
400	0.56	0.63	0.94
500	0.40	0.44	0.67
600	0.24	0.26	0.40
700	0.08	0.08	0.12
800	0.06	0.06	0.11
900	0.05	0.05	0.08
1000	0.03	0.03	0.05
1100	0.02	0.02	0.03
1200	0	0	0

Or, it can be expressed as:

$$\eta_{fi} = \frac{E_{fi,d}}{E_d} = \frac{G_k + \psi_{fi}Q_{k,1}}{\gamma_G G_k + \gamma_{Q,1} Q_{k,1}} \tag{6.3}$$

where

E_d is the design value for normal temperature design;
G_k is the characteristic value of a permanent action;
$Q_{k,1}$ is the characteristic value of the leading variable action 1;
γ_G is the partial factor for permanent action, taken as 1.35;
$\gamma_{Q,1}$ is the partial factor for variable action, taken as 1.50; and
ψ_{fi} is the combination factor for a fire situation, given by $\psi_{1,1}$ or $\psi_{2,1}$
according to EN 1991-1-2, as shown in Table 6.4.

This value is more conservative and is used in simplified designs.

The reduction factor changes with the load ratio. For simplification, a value of 0.65 is recommended for the reduction factor.

6.5 FIRE DESIGN RESISTANCE

Composite columns always have a higher fire resistance than steel columns, taking advantage of the thermal properties of concrete. In addition, the reinforcement also contributes to fire resistance. Composite columns made

Table 6.4 Values ψ of for buildings

Action	$\psi_{1,1}$	$\psi_{2,1}$
Domestic, residential area	0.5	0.3
Office area	0.5	0.3
Congregation area	0.7	0.6
Shopping area	0.7	0.6
Storage area	0.9	0.8
Traffic area, vehicle $\leq$ 30 kN	0.7	0.6
Traffic area, 30 kN $\leq$ vehicle $\leq$ 160 kN	0.5	0.3
Roofs	0.0	0.0
Snow loads, H $\geq$ 1000 m above sea level	0.5	0.2
Snow loads, H < 1000 m	0.2	0.0
Wind loads	0.2	0.0

of totally encased steel sections (SRC) are covered only by the tabular and advanced calculation methods in accordance with EN 1994-1-2. Simple calculation models are applicable only for composite columns made of partially encased steel sections and concrete-filled hollow sections. The fire resistance of composite columns with fully concrete-encased steel sections may be treated in the same way as reinforced concrete columns.

6.5.1 Tabular data

The fire resistance of fully concrete-encased steel sections is determined by the thickness of the concrete cover of the steel section and the reinforcement. The concrete cover permits the steel to retain good protection of its strength. Table 6.5 gives the minimum cross-sectional dimensions, minimum concrete cover of the steel section, and minimum axis distance of the reinforcing bars of fully concrete-encased steel sections. The design table is valid for braced frames subjected to concentric and eccentric loads.

Using the minimum requirements in Table 6.5, the SRC column can achieve the required fire resistance irrespective of the load level in the fire conditions. For example, for a fire resistance time of 90 minutes (R90), which can be achieved by a cross-section with dimensions h_c and b_c of at least 250 mm, the minimum concrete cover to the steel section and the reinforcement are 40 and 20 mm, respectively, in accordance with EN 1994-1-2.

For SRC columns in a fire situation, minimum of four bars with a minimum diameter of 12 mm should be placed in the corners of the

Table 6.5 Minimum cross-sectional dimensions, minimum concrete cover of the steel section, and minimum axis distance of the reinforcing bars of composite columns made of fully encased steel sections

	Standard fire resistance					
	R30	R60	R90	R120	R180	R240
Minimum dimensions h_c and b_c (mm)	150	180	220	300	350	400
Minimum concrete cover of steel section c (mm)	40	50	50	75	75	75
Minimum axis distance of reinforcing bars u_s (mm)	20*	30	30	40	50	50
Or						
Minimum dimensions h_c and b_c (mm)	–	200	250	350	400	–
Minimum concrete cover of steel section c (mm)	–	40	40	50	60	–
Minimum axis distance of reinforcing bars u_s (mm)	–	20*	20*	30	40	–

*These values have to be checked according to EN 1992-1-1.

cross-section. The minimum and maximum amounts of longitudinal reinforcement should fulfill the normal design requirements of EN 1994-1-1. The reinforcing links (stirrups) should fulfill the normal design requirements of EN 1992-1-1. These requirements are needed to reduce the danger of spalling of the concrete under fire exposure.

If concrete has only an insulating function, a simplified Table 6.6 may be used for fire design. In this case, the minimum concrete cover to the steel section is reduced. For SRC column design, the fire resistance R30 to R180 may be fulfilled when a concrete cover c to the steel section satisfies the requirements of Table 6.6. A further requirement is that the steel fabric reinforcement with a minimum diameter 4 mm and a maximum spacing 250 mm should be placed in the cross-section to prevent the spalling of concrete during a fire.

Table 6.6 Minimum concrete cover for a steel section with concrete acting as fire protection

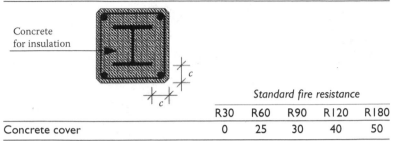

Concrete
for insulation

	Standard fire resistance				
	R30	R60	R90	R120	R180
Concrete cover	0	25	30	40	50

6.5.2 Advanced calculation models

The advanced calculation models are applicable for individual members, subassemblies, and entire structures. Compared to tabulated data and simple calculation models, advanced calculation models give an improved approximation of the actual performance under fire conditions. As mentioned previously, simple calculation models are applicable only for composite columns made of partially encased steel sections and concrete-filled hollow sections. However, advanced calculation models can be used for any type of cross-section.

Advanced calculation models may cover the thermal response of the structure to any temperature–time heating curve, and shall be based on the acknowledged principles of heat transfer. The effects of nonuniform thermal exposure and heat transfer to adjacent components may be considered in the advanced calculation models. The advanced calculation models may be accomplished by finite-element analysis software (e.g., ABAQUS) and shall be verified by relevant tests.

Bibliography

ABAQUS 2014. ABAQUS Standard User's Manual 6.14.

AISC 2010. Specification for Structural Steel Buildings. American Institute for Steel Construction. ANSI/AISC 360-10. Chicago, Illinois.

Anslijn R. and Janss J. 1974. "Le calcul de charges ultimes des colonnes metalliques enrobes de beton," C.R.I.F. Report MT89, Brussels.

Chen C., Astaneh-Asl A., and Moehle J.P. 1992. "Behavior and design of high strength composite columns," *Structures Congress '92*, ASCE, San Antonio, Texas, 820–823.

Chen C.C. and Lin N.J. 2006. "Analytical model for predicting axial capacity and behaviour of concrete encased steel composite stub columns," *Journal of Constructional Steel Research*, 62(5): 424–433.

Chen C.C. and Yeh S.C. 1996. "Ultimate strength of concrete encased steel composite columns," *Proceedings of the Third National Conference on Structural Engineering*, 2197–2206.

Chiew S.P. and Cai Y.Q. 2014. *Design of Composite Steel and Concrete Structures with Worked Examples to Eurocode 4*. Research Publishing, Singapore.

Comite Euro-International Du Beton (CEB) 1990. *CEB-FIP Model Code 1990*. Thomas Telford, London.

Cusson D. and Paultre P. 1995. "Stress-strain model for confined high-strength concrete," *Journal of Structural Engineering*, 121(3): 468–477.

EC2 2004. *Eurocode 2: Design of Concrete Structure Part 1-1: General Rules and Rules for Buildings*. London, British Standards Institution, BS EN 1992-1-1.

EC3 2005. *Eurocode 3: Design of Steel Structures Part 1-1 General Rules and Rules for Buildings*. London, British Standards Institution, BS EN 1993-1-1.

EC3 2007. *Eurocode 3: Design of Steel Structures Part 1-12 Additional Rules for Extension of EN 1993 up to Steel Grades S700*. London, British Standards Institution, BS EN 1993-1-12.

EC4 2004. *Eurocode 4: Design of Composite Steel and Concrete Structure Part 1-1: General Rules and Rules for Buildings*. London, British Standards Institution, BS EN 1994-1-1.

EC8 2004. *Eurocode 8: Design of Structures for Earthquake Resistance Part 1 General Rules, Seismic Actions and Rules for Buildings.* London, British Standards Institution, BS EN 1998-1-1.

El-Tawil S. 1996. "Inelastic Dynamic Analysis of Mixed Steel-Concrete Space Frames," *UMI Dissertation Services,* Michigan.

El-Tawil S. and Deierlein G.G. 1999. "Strength and ductility of concrete encased composite columns," *Journal of Structural Engineering,* 125(9): 1009–1019.

Ellobody E. and Young B. 2011a. "Eccentrically loaded concrete encased steel composite columns," *Thin-Walled Structures,* 49(1): 53–65.

Ellobody E. and Young B. 2011b. "Numerical simulation of concrete encased steel composite columns," *Journal of Construction Steel Research,* 67(2): 211–222.

Falkner H., Gerritzen D., and Jungwirth D. 2008. "The new reinforcement system: Compression members with SAS 670 high strength reinforcement steel," *Beton- und Stahlbetonbau,* 103(5): 1–34.

fib 2010. *fib Model Code for Concrete Structures 2010.* Ernst & Sohn, Berlin, Germany.

Goncalves R. and Carvalho J. 2014. "An efficient geometrically exact beam element for composite columns and its application to concrete encased steel I-sections," *Engineering Structures,* 75(5): 213–224.

Han D.J. and Kim K.S. 1995. "A study on the strength and hysteretic characteristics of steel reinforced concrete column," *Journal of the Architectural Institute of Korea,* 11(4): 183–190.

Han, D.J., Kim P.J., and Kim K.S. 1992. "The influence of hoop bar on the compressive strength of short steel reinforced concrete columns," *Journal of the Architectural Institute of Korea,* 12(1): 335–338.

Hoang T.T. 2009. "Seismic Behavior of Steel Reinforced Concrete Columns with Axial Compressive Force," *Master Thesis,* National Taiwan University of Science and Technology, Taipei.

Johnson R.P. 2004. *Composite Structures of Steel and Concrete* (third edition). Blackwell Scientific Publication.

Johnson R.P. and Anderson D. 2004. *Designers' Guide to EN 1994-1-1: Eurocode 4, Design of Composite Steel and Concrete Structures. Part 1-1, General Rules and Rules for Buildings.* London, Thomas Telford.

Kim C.S., Park H.G., and Chung K.S. 2012. "Eccentric axial load testing for concrete encased steel columns using 800 MPa steel and 100 MPa concrete," *Journal of Structural Engineering,* 138(8): 1019–1031.

Kim C.S., Park H.G., and Chung K.S. 2014. "Eccentric axial load capacity of high-strength-steel-concrete composite columns of various sectional shapes," *Journal of Structural Engineering,* 140(4): 04013091-1-12.

Kim D.K. 2005. "A Database for Composite Columns," *MSc Thesis,* Georgia Institute of Technology, Atlanta, Georgia.

Legeron F. and Paultre P. 2003. "Uniaxial confinement model for normal and high-strength concrete column," *Journal of Structural Engineering,* 129(2): 241–252.

Li, V.C., Wang S., and Wu C. 2001. "Tensile strain-hardening behavior of polyvinyl alcohol engineered cementitious composite (PVA-ECC)," *ACI Materials Journal*, 98(6): 483–492.

Liew J.Y.R. and Xiong M.X. 2015. *Design Guide for Concrete Filled Tubular Members with High Strength Materials to Eurocode 4*. Singapore, Research Publishing.

Mander J.B., Priestley M.J.N., and Park R. 1988. "Theoretical stress–strain model for confined concrete," *Journal of Structural Engineering*, 114(8): 1804–1826.

Matsui C. 1979. "Study on elasto-plastic behaviour of concrete-encased columns subjected to eccentric axial thrust," *Annual Assembly of Architectural Institute of Japan*, 1627–1628.

Mirza S.A., Hyttinen V., and Hyttinen E. 1996. "Physical tests and analyses of composite steel-concrete beam–columns," *Journal of Structural Engineering*, 122(11): 1317–1326.

Mirza S.A. and Lacroix E.A. 2004. "Comparative strength analyses of concrete-encased steel composite columns," *Journal of Structural Engineering*, 130(12): 1941–1953.

Mirza S.A. and Skrabek B.W. 1992. "Statistical analysis of slender composite beam–column strength," *Journal of Structural Engineering*, 118(5): 1312–1331.

Roik K. and Diekmann C. 1989. "Experimental studies on composite columns encased in concrete following loading," *Der Stahlbau*, 58(6): 161–164.

Ruesch H. et al. 1983. *Creep and Shrinkage*. Springer-Verlag.

Shanmugam N.E. and Lakshmi B. 2001. "State of the art report on steel-concrete composite columns," *Journal of Constructional Steel Research*, 57(10): 1041–1080.

Sheikh S.A. and Uzumeri S.M. 1980. "Strength and ductility of tied concrete columns," *Journal of the Structural Division* 106 (ASCE 15388 Proceedings), 106(5): 1079–1102.

SSEDTA project 2002. Structural Steelwork Eurocodes, Development of a Trans-National Approach.

SSRC Task Group 20 1979. "A specification for the design of steel-concrete composite columns," *AISC Engineering Journal*, 16: 101–115.

Structural & Conveyance Business, Design Guide for Concrete Filled Columns. Corus Tubes. 2002.

Stevens R.F. 1965. "Encased stanchions," *Structural Engineering*, 43(2): 59–66.

Tsai K.C., Lien Y., and Chen C.C. 1996. "Behaviour of axially loaded steel reinforced concrete columns," *Journal of the Chinese Institute of Civil and Hydraulic Engineering*, 8(4): 535–545.

TW-SRC Code 2004. *Design Code and Commentary for Steel Reinforced Concrete Structures (in Chinese)*. Taipei, Taiwan Construction and Planning Agency.

Weng C.C. and Yen S.I. 2002. "Comparisons of concrete-encased composite column strength provisions of ACI code and AISC specification," *Engineering Structures*, 24(1): 59–72.

Wight J. and MacGregor J. 2012. *Reinforced Concrete, Mechanics & Design* (sixth edition). Pearson, New Jersey.

Index

Advanced calculation models, 73, 81
Axial compression
   resistance of cross-section, 50, 51
   resistance of members, 50–54

British Standard European Norm
   206–1 (BS EN 206–1), 7
Buckling curves and member
   imperfections for SRC
   columns, 51, 52
Building and Construction Authority
   (BCA), 8

Calcareous concrete, 75
CEB Model Code 90, 33
Certification, 7
CFT, *see* Concrete filled steel tubes
Composite columns, 1, 78–79
Compression and bending
   combination, 54
   resistance of cross-section, 54–61
   resistance of members in
      combined compression and
      biaxial bending, 65–67
   resistance of members in
      combined compression and
      uniaxial bending, 64–65
   second-order effects and member
      imperfections, 61–64
Compression resistance, 57
Compressive strain, 6
Concrete, 7, 28, 75–77
   core, 18

material, 74
   time-dependent stress–strain
      curves, 34
   total strain, 34
Concrete confinement model, 19–20
   for HCC, 27–29
   modified confinement model for
      SRC columns, 29–32
   for PCC, 20–27
Concrete creep and shrinkage model
   Eurocode 2, 39–42
   fib Model Code 2010, 33–39
   load redistribution, 42–45
   in steel-reinforced concrete
      columns, 45
Concrete filled steel tubes (CFT), 1–2
Confinement effect, 18, 29–30, 31
Confinement model for HCC, 27
   lateral confining stress from steel
      section, 28–29
Confinement model for PCC, 20
   Eurocode2 model, 24–26
   Fédération internationale du beton
      model code 2010, 26–27
   Legeron and Paultre model, 23–24
   Mander model, 20–23
Confining effect, 26
Confining stress for HCC, 27
Constant stress, 35
Construction process, 3
Creep
   coefficient, 34–35
   Eurocode 2, 39–41